AEROTOWING GLIDERS

John Marriott

authorHOUSE®

AuthorHouse™ UK Ltd.
500 Avebury Boulevard
Central Milton Keynes, MK9 2BE
www.authorhouse.co.uk
Phone: 08001974150

These notes have been compiled in the interest of safety and operational efficiency, using the established practices of a number of gliding organizations and the experience of very skilled tug pilots. They are offered to club tugmasters, chief tow-pilots, and glider tug pilots as a source of guidance and recognized good practices. I would welcome feedback, with a view to updating the book in the future, to johnPmarriott@gmail.com

The author accepts no responsibility for any of the suggested practices contained in this document. Ultimately, always use your own best judgement. The notes are intended as a general guide to glider-tugging operations.

First published by AuthorHouse 05/25/2011

ISBN: 978-1-4567-7515-5 (sc)
ISBN: 978-1-4567-7516-2 (e)

Contents

Foreword

by
Sir John Allison, KCB, CBE,
President of Europe Air Sports

I first met John Marriott when we were both serving at Royal Air Force Wildenrath, in Germany in the early 1980s and were fellow members of the Phoenix Gliding Club. John stood out then by virtue of his huge enthusiasm for flying in general and gliding in particular, coupled with the serious way that he applied himself. This combination of qualities has served him well while he has pursued a successful career as an airline pilot. The same attributes made him a very good gliding instructor, even as a young man, and I was entirely comfortable when he taught my two teenage children to glide.

John retains that enthusiasm, coupled with personal altruism, a desire to serve the gliding movement in general, and the cause of flight safety in particular. The outcome is this excellent book, which is produced at his expense and the profits of which will be given to the Air League. Between its pages we are given another gift – the distillation of his decades of experience as a tug pilot and gliding instructor.

All forms of aviation are inherently dangerous, and when things go wrong in the air the situation can go from normal to life threatening with terrifying speed. This is never truer than in the seemingly routine operation of towing gliders, which is simply a consequence of the laws of physics. Aerotowing, as with everything in flying, is rendered acceptably safe by self discipline in the conscientious and thorough application of hard-won experience from the past. In his book, John offers us that experience pre-packaged and clearly expressed. I most warmly commend the book to anyone who is, or aspires to be, a glider-tug pilot.

John Allison

Acknowledgements

I have drawn freely on good practice guidelines developed over many years by the gliding community, and I'm greatly indebted to all those individuals who have contributed. Much of the content has previously been published in one form or another by individuals, organizations, and clubs. I acknowledge the work done by others that has helped so much in the preparation of this reference book. There are too many examples to acknowledge everyone individually, but I would like to specifically thank Steve Longland for his permission to publish his excellent gliding and aerotowing illustrations and also thanks to my wife Anna for her support. I would also like to thank the British Gliding Association (BGA) and its band of volunteers. My thanks go to Paul Cooper, Dave Smith, Ron Smith, Ted Norman, Tez Bowler, Fergal Goodman, Paul Johnson (FlightlineUK), Peter Atkinson, Myles Noton, and Guy Westgate of the Swift Aerobatic Display Team for some great photos. Guy is an advanced aerobatic display pilot. Copying him and his team inverted on aero tow should not be attempted! Special thanks to Peter Atkinson for a great front cover.

Illustrations from the Civil Aviation Authority of the UK and Federal Aviation Administration of the USA have been used to promote safety, as they were intended by those organizations. Extracts from the excellent UK Confidential Human factors Incident Reporting Programme (CHIRP) are gratefully reproduced. I thank Master CFI Jon Cooke of the UK's Light Aircraft Association (LAA) who contributed his expertise. Anecdotes are included throughout. When they wanted to be, I have credited contributors at the end of the borrowed or quoted passage. I thank them for their valuable contributions. I would like to re-emphasize that I don't consider myself to be the author of this book so much as the compiler of information, which I hope will result in safer, more efficient, and more fun glider aerotowing. Any proceeds from the sales of this book will be donated to the Air League of Great Britain, which promotes aviation for young people.

Introduction

This book on aerotowing gliders was written because there is little reference material published about the subject worldwide. The best I have found is *Towplane Manual* by Burt Compton and published as part of Bob Wander's Gliding Mentor Series in the USA. So, because of the lack of published information, I thought it important to gather the wealth of knowledge that is out there on the subject, collate it, and present it to our community in the interests of safety and efficiency

This book is intended as a comprehensive guide to glider-towing operations, with that all important emphasis on safety. The intent is to provide all the relevant information in one straightforward, easy-to-read book. The notes are intended to be very generic and non-country specific. Even though local procedures differ, hopefully the information should be useful to any glider tug pilot, anywhere in the world. Each gliding organization has its operating environment and problems, therefore should adapt, further, or improve these suggestions to suit their own needs.

You will find that some important points are emphasized and sometimes repeated.

It is fundamental that every tug pilot be a person who is both trustworthy and highly reliable, because tug piloting is a flying task with huge responsibility placed on the pilot. Aerotowing is expensive, can be noisy, and has its own special hazards. These factors have a bearing on the very existence of gliding, and it is therefore essential that glider aerotowing be carried out safely, efficiently, thoughtfully, and paying particular regard to neighbours. Your particular aerotowing should, of course, be carried out in accordance with national laws, regulations, and procedures and in conjunction with your organization's flying rules.

As the pilot in command of an aircraft, you are ultimately responsible for the safe conduct of the flight and the actions that you choose to take. The glider pilot's requirements should, of course, be accommodated as far as possible.

Glider aerotowing should be good for your general flying skills. During my more than 25 years as a flying and gliding instructor, I have noticed that most glider-tug pilots are often also glider pilots and have above average handling and situational awareness skills. Flying tugs should of course also be quite good fun!

I hope that this comprehensive book will meet the ground-school requirements of any current or future glider-towing ratings.

In writing this book, I was faced with the reality that what we call here in Great Britain a 'tug pilot' will go by a variety of different terms in other countries, ranging from tow pilot, tow plane pilot, tuggie, and probably a few more. So, in the tradition of true British compromise I will hereafter call him or her a 'tug pilot'. As a Brit, I've also opted for UK spelling; so, if you are reading this in North America, there might appear to be some odd spellings.!

Section 1 - Normal Procedures

Safety

This book must start with the most important aspect of any flying: safety. Over the years, many accidents and incidents have been categorized as *Pilot Error*. This term is too simplistic and does little to prevent a similar thing happening again. I would like to introduce you to the terms *Fair Blame* and *Just Safety Culture*. It has been learned in professional aviation and other safety-critical industries that blaming an individual is often unfair and does not advance the practice and doctrine of safety.

Simply throwing the book at an individual is not the most constructive approach – after all, nobody goes out with the intention of having an accident. Therefore, in order to encourage openness it was decided to establish a 'No Blame Culture'. This didn't work because it is not possible to simply have 'no blame'. For example, blatant breach of the law or downright recklessness cannot be overlooked. However, a Fair Blame environment, in which if you make an honest mistake and put your hands up to it you will not be punished. A witch hunt in order to blame an individual serves no use when often the deeper problem is systemic or due to other factors. 'To err is human', and we are all capable of making mistakes. So, let us not blame, but let us understand.

Full of Holes

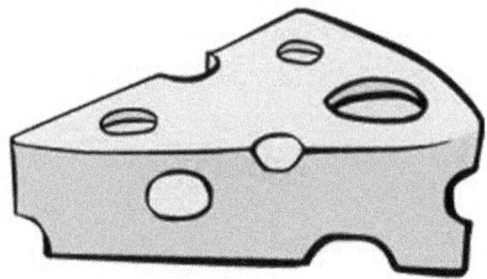

Professor James Reason of Manchester University introduced the 'Swiss Cheese Analogy'. This says there are usually many defenses in place to avert an accident, but each defense is not perfect and has gaps, weaknesses, or flaws. Some of these defences could include:

- Use of Checklists
- Rules
- Training
- Proficiency
- Skill
- Knowledge
- Flying Currency
- Standard Operating Procedures

Twist the Swiss cheeses around, and eventually these holes will line up and an accident will occur. *We must strive to prevent the Swiss cheese holes from lining up.*

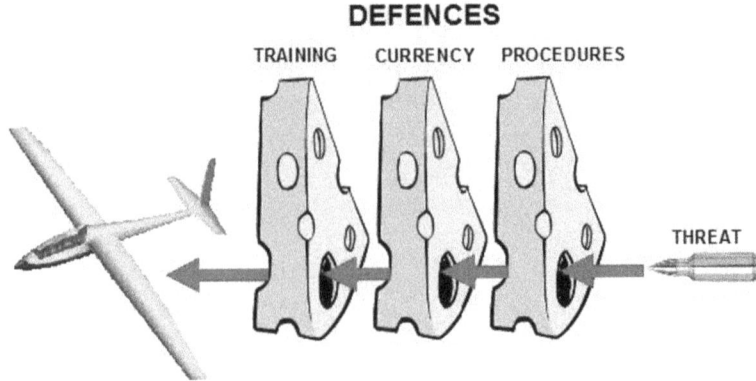

If you are involved in an accident or incident, may I encourage you to report it, honestly, without delay and confidentially, if you so desire. Let others learn from your experience. The military have been known to call this 'safety in knowledge'.

What's It All About?

Having emphasized the obligation to safety, at the risk of stating the obvious, I point out that aerotowing operations exist to provide launches to glider pilots. This awareness of service means, for example, launch waiting times should be minimized and the needs of the glider pilot met. Soaring pilots may require you to drop them in the nearest lift source, whilst a student might need to remain near the field with gentle manoeuvres.

In many cases, the organization's or club's tug is its most valuable asset. The insurance and servicing costs can also eat into a substantial part of the business funds. It is therefore important to minimize cost to the club and thereby to the individual. This can be achieved by:

- Conserving the engine by careful handling
- Reducing fuel burn by accurate flying
- Minimizing ground running time

Achieving all this is also immensely rewarding, and good tug pilots normally display the qualities of wanting to constantly improve their skills and better their performance. Organizations should strive to utilize pilots who take a professional approach to the task.

In short, the three main points that a tug pilot should always aim to achieve are:

- SAFETY – Safe flying and good airmanship should be expected at all times and are everyone's responsibility
- ACCURACY – Delivering the best service to the glider pilot
- EFFICIENCY – Handling the airframe and engine with precision, optimizing fuel economy

What Makes a Good Glider Aero-Towing Pilot?

The best candidates for future tug pilots are normally glider pilots who also hold a power pilot's licence. Glider pilots understand gliding operations, and so are one step ahead before they start. However, there is no reason a non-glider pilot cannot become a tug pilot – it's up to each organization to consider their own requirements. In some countries, a specific rating is required. However, if no formal rating is required, I recommend that gliding organizations at least give a potential tug pilot with no gliding experience some sort of introduction to glider flying before training them to tug.

Good general stick and rudder skills are required. These, coupled with a commercial awareness to provide the glider pilot with a high-quality service, will save the club money by looking after the tug and conserving expensive fuel. The most important ingredient to being a tug pilot is undoubtedly safety – it's really important to be conscientious, have excellent situational awareness and keep a really good lookout at all times – both in the air and on the ground.

For example, a good tug pilot might notice a ground handler who is attempting to put the tow rope on the winch launch hook by mistake, or perhaps the pilot will see a conflicting glider on approach when the tug is about to take off.

Another example – for better situational awareness I was taught to be a **NUTA**:

- **N**otice – The wind is 290/15kts
- **U**nderstand – That means a 15 kt cross wind and little into wind component
- **T**hink **A**head – What effect is this going to have on me towing today and what about any emergency?

I know of some tug pilots who just wanted to build hours towards a commercial pilot's licence but were subsequently bitten by the gliding bug and became ardent glider pilots, so that's really gratifying to see.

As each club will have its own tug type, site, and specific operating problems, the site training of the new tug pilot should be up to each club's tugmaster.

Don't let an aircraft take you someplace your brain didn't visit five minutes earlier.

An old, wise, airline training captain

Club Tug Master or Chief Tug Pilot

The Chief Flying (Gliding) Instructor is not always a power pilot, therefore a separate appointment is sometimes necessary and appropriate. The chief tug pilot is responsible to his or her gliding organization for towing operations. This normally includes training of tug pilots, maintenance and airworthiness of tug aircraft, making up/serviceability of tow ropes, and perhaps scheduling of tug-pilot duties. There might be a deputy chief tuggie to assist.

Thus, the tug master needs to be a versatile individual who can cope not just with the aerotowing itself but also to take responsibility for tug maintenance, tow ropes, pilot rosters, personalities etc. It can be very difficult for the chief tug pilot to monitor the validity of every tug pilot's licence, ratings, medical certificate, and other certifications, so these things should be the responsibility of the individual to manage.

Approval to Tow Gliders

It might be interesting to note that, apart from an appropriate pilot's licence, at the time of publication (January 2011) there is no minimum experience or specific qualifications for becoming a club tug pilot in the UK. This is normally left to the discretion of the chief tug pilot. It's also interesting to note that this non-regulatory approach has served us well for many years. In the UK, there is an operational regulation that the sum of the tows made by the tug pilot and the glider pilot, in their respective capacities, shall not be less than six – not a bad idea.

In other countries around the world, different rules apply, and often a formal rating or qualification is required – check with your regulatory authority. The chief tug pilot normally has organizational authority to revoke or suspend any club tug pilot's approval for any reason, but if the issue involves errors in flying, my experience is that it is normally better to offer training and assistance to the pilot rather than taking punitive action. People do learn from their mistakes. As I suggested earlier, it is the pilot's responsibility to maintain a current licence, medical, ratings, and certificate of experience. The tug pilot would also normally be a full flying member of that club or organization and hopefully also a glider pilot. At the time of writing there are no known 'professional' tug pilots in the UK. This of course differs considerably around the world.

A new tug pilot with previous towing experience would normally receive a thorough briefing and perhaps also have completed an instructional introductory flight with the chief tug pilot or one of his deputies before flying tug aircraft at the new club or organization. Local procedures, noise abatement, commercial awareness, and of course safety should be emphasized.

Currency, Rest, Training, and Checks

It is probably impractical to put a limit for time spent in between flying tug sorties. This could vary from individual to individual due to experience, competence level, recent experience in similar types, and other factors. The best person to make such a judgement is the pilot himself or the chief tug pilot. As a guide, the British Gliding Association has produced a very good 'barometer' as an aid to deciding gliding currency.

Maintaining currency is important, but being well rested and fit are equally important. As with many things associated with aviation there is a balance. Here is an aviator's acronym to help you decide if you are okay to tow gliders – **I'M SAFE**:

- **I** – Illness (any symptoms?)
- **M** – Medication (your doctor may not know you are a pilot)
- **S** – Stress (upset following an argument maybe?)
- **A** – Alcohol/Drugs (obvious, I hope!)
- **F** – Fatigue (good night's sleep or too much flying)
- **E** – Eating (to keep correct blood-sugar level)

Additional considerations for the glider aerotowing pilot could include:

- Don't aerotow for too long without a break
- If you are towing on a daily basis, have the odd day or two off
- Wear a suitable hat but one that doesn't impair your vision – without a big peak for example
- Stay hydrated (that's drinking lots of water, not beer!)

SAFE FLYING!

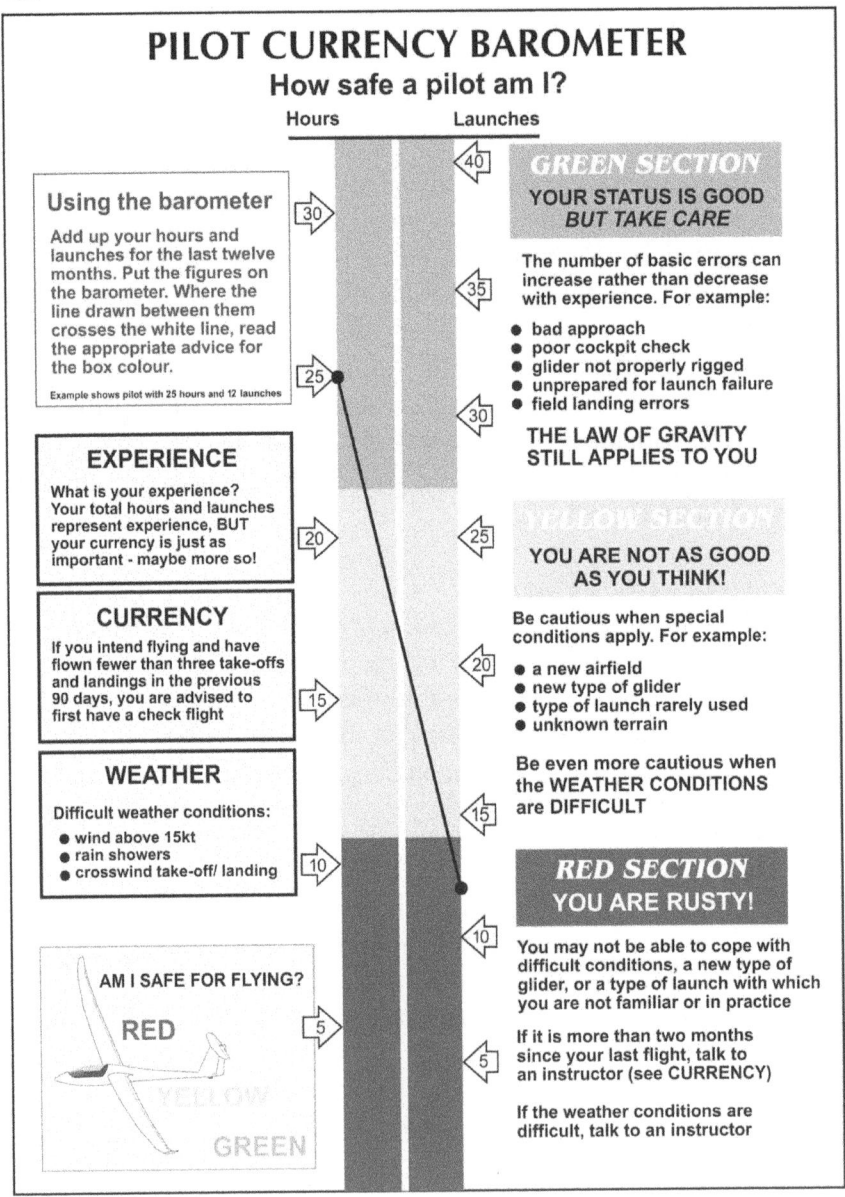

PILOT CURRENCY BAROMETER
How safe a pilot am I?

Hours   Launches

Using the barometer

Add up your hours and launches for the last twelve months. Put the figures on the barometer. Where the line drawn between them crosses the white line, read the appropriate advice for the box colour.

Example shows pilot with 25 hours and 12 launches

EXPERIENCE

What is your experience? Your total hours and launches represent experience, BUT your currency is just as important - maybe more so!

CURRENCY

If you intend flying and have flown fewer than three take-offs and landings in the previous 90 days, you are advised to first have a check flight

WEATHER

Difficult weather conditions:

- wind above 15kt
- rain showers
- crosswind take-off/ landing

AM I SAFE FOR FLYING?

RED

YELLOW

GREEN

GREEN SECTION
YOUR STATUS IS GOOD
BUT TAKE CARE

The number of basic errors can increase rather than decrease with experience. For example:

- bad approach
- poor cockpit check
- glider not properly rigged
- unprepared for launch failure
- field landing errors

THE LAW OF GRAVITY STILL APPLIES TO YOU

YELLOW SECTION

YOU ARE NOT AS GOOD AS YOU THINK!

Be cautious when special conditions apply. For example:

- a new airfield
- new type of glider
- type of launch rarely used
- unknown terrain

Be even more cautious when the WEATHER CONDITIONS are DIFFICULT

RED SECTION
YOU ARE RUSTY!

You may not be able to cope with difficult conditions, a new type of glider, or a type of launch with which you are not familiar or in practice

If it is more than two months since your last flight, talk to an instructor (see CURRENCY)

If the weather conditions are difficult, talk to an instructor

(This diagram is obviously supposed to be in colour but I hope it doesn't take too much imagination to interpret it in various shades of grey)

After a few days towing at a competition, I gradually developed a pain in my right ankle which was bruised and swollen. I think this was due to sustained pressure on the rudder bar attempting to stay in balance, causing a repetitive strain injury.

I suggest that each tug pilot should have an occasional instructional flight with the chief tug pilot or nominated deputy to include normal tows, general handling, and emergency drills. The flight should be viewed as an opportunity to improve flying skills, and practice unusual situations – it should not be considered a test. Further and recurrent training could include attending safety seminars and such like, which are also of great benefit. This recurrent training should be flexible and suit the needs of the individual pilot. For example, a new tug pilot will need more support than someone with thousands of hours and who has been glider towing for years, but of course, even the most experienced pilot never stops learning.

Authorization and Responsibilities

Most tug pilots would be self-authorizing for local glider aerotowing, but it is usually a legal requirement that all powered aircraft movements are recorded. A movement log should be kept, and all aero-tow retrieves, positioning flights, or hire flights should be entered. Special authorisation is often required for off home airfield glider retrieves. For safety reasons, particular attention should be paid to overdue procedures. Check your insurance to ensure you are actually covered for all activities you intend to engage in. For example, field retrieves might require a special clause.

Meet Your Aircraft

Care must be taken when moving aircraft by hand, not only to avoid pushing them into things or being damaging them by people pushing on the wrong places, but crucially, because propellers must always be treated as live, even when the magneto switches are off.

Some aircraft, the Husky for example, should have aileron spades covered with padding to prevent injury, and some motorgliders will need the tail-wheel castering before they can be moved in a confined space.

I find it easier to push most aircraft types backward, against the wing leading edge, rather than forward, against the weak trailing edge. The exceptions could be aircraft with strong wing struts that could easily be pulled forward. For safety reasons, I strongly advise not to move aircraft by their propellers.

I know of an unfortunate chap many years ago who lost his legs pulling an aircraft out of a hangar in Germany, when the engine unexpectedly fired. Others have lost their lives.

Propeller Swinging

Here we come to an extremely dangerous bit – avoid doing it if possible!

If the aircraft simply has a flat battery, consider charging it as the first preference or, in extremis, jump starting the machine. But be aware as jump starting is also potentially dangerous. If swinging is carried out, it should be done with great caution, and it is imperative that you have had training in the techniques. Some engines such as the Lycoming can be difficult to hand swing, but Gipsy Major and similar in-line engines are more straightforward. Rotax and other belt-driven engines should not be hand-swung.

As stated, propeller swinging is best avoided if possible. If you are going to do it, it's easier on an engine of small displacement, low-compression ratio, and with a two-bladed propeller.

A few general safety precautions could include:

- Always treat a propeller as live
- Ensure the aircraft brakes are set to on
- Ensure that there is a competent person in the cockpit
- Ensure that the area around the propeller is clear of oil, water, or anything that could cause you to slip into the rotating propeller
- When swinging, adopt a stance such that you tend to lean away from and swing away from the propeller
- Establish clear and robust communications
- Wear no loose clothing but consider wearing gloves
- Never hand swing an aircraft with a four-bladed propeller as the blades are so close together you are in danger of being hit by the following blade when the engine fires
- Ensure chocks are in place
- Ensure that the throttle is set correctly for engine start

Important communication terms such as *contact* and *off* clearly distinguish between terms that can easily be misheard such as 'on' and 'off'. It is also fairly common practice for a *thumb up* to mean contact and *thumb down* to mean off.

Bear in mind that four-bladed propellers are lighter as they are normally made of wood and have less inertia to carry through the compression. They are usually smaller in diameter, so more force is required and the next blade comes around in half the time.

If in doubt – don't!

I've heard of a site where they always have to swing the propeller as the aircraft doesn't have a starter. They fitted a fixed anchor point and a short length of wire with rings. While the propeller is swung the aircraft is attached to the anchor through the normal release. Once in the cockpit with the engine running, the pilot/prop swinger can release and simply taxi to the launch point.

I consider this be a reasonable and acceptable risk assessment.

Tow Ropes

There are many different types of rope, and I don't intend to go into great detail here. Suffice to say they each have their advantages and disadvantages. An ideal rope length for normal club operations is about 160 ft to 180 ft (45 m to 55 m). Shorter ropes are sometimes used for field retrieves but make it harder for the glider pilot to remain in position behind the tug. Long ropes normally make it easier for the glider pilot on long tows. Extra-long tow ropes or two shackled firmly together can sometimes be easier for the glider pilot but can have associated problems.

When we first started training the 'Roll-on-tow' manoeuvre in an MDM-1 Fox, we trialled a 96m rope (twice the usual length) – our rationale was that errors in rolling would produce a smaller angular displacement at the tug end (An Extra300L). What we found, however, was that in level flight when there was not much tension on the rope, the rope was far too long. Not only did the rope hang down creating its own bow, but any movements from either the tug or glider would travel up and down the rope as a standing wave. We quickly changed back to a 49m rope for display flying and all training.

Guy Westgate
Swift Aerobatic Display Team

A weak link must be fitted to the tug end of the rope so that if it snags on landing it will not subsequently pull the tug into the ground or rip its tail off! Some organizations have another weaker weak link fitted at the glider end of the rope. Blue Tost blue weak links are often used for towing single or dual gliders; they have breaking strains of about 1348 lbs (600 deca Newtons for our European friends). Tug flight manuals should specify the max weak link load for a specific tug aircraft, for example, the Super Cub is 1,000 lbs. Yellow, white, or blue Tost links are however quite commonly used, but operators, clubs, and tug masters should check their tug-flight manuals to ensure that the links in use at the club do not exceed that permitted by the tug-flight manual. Most glider-flight manuals also give maximum aerotow weak link strengths, and so the link strength must not exceed the lower of that permitted by either the tug or glider-flight manuals.

During the daily check, look for damage to the rope or embedded knots. This can be done by running the rope carefully through your hands. Check that the rings are symmetrical and free from distortion or damage. The big ring (if fitted) is designed to bend under stress before the small ring. This is to prevent the smaller ring from getting jammed in the release mechanism if it becomes distorted.

I have heard of a club that had a problem with a Dynema rope that was put away wet on-board an aircraft (wound in a figure of eight). Winter subzero overnight temperatures caused ice crystals that damaged the rope when it was unwound the following morning. Message: keep wet ropes from freezing.

Unserviceable ropes should not be returned to the normal stowage but left with a note or label as to the defect. Repairing a broken rope or changing a broken weak link is a specialized job. Please don't try it unless you know what you are doing. If in doubt, take a new rope from your rope store.

To check that the release operates correctly, attach the rope, pull with a reasonable force, and have a competent person operate the release from the cockpit. Releasing the rope from the hook end is not sufficient because it does not check the full system. Health and safety note – don't fall over!

The tow rope should be laid out behind the tug in such a way that the risks of vehicles driving over it or aircraft taxiing over it are minimized. Laying out the tow rope in an orderly manner also reduces the risk of it becoming tangled when taxiing off and reduces the chances of it becoming entangled in the propeller.

Retractable Tow Ropes

If your aircraft has a retractable rope system, the weak link is normally placed at the glider end to protect it during the tow. Therefore, if on the approach the rope is left extended, either by forgetfulness or by technical defect, the towing aircraft will have no weak-link protection. Depending on the aircraft and retraction system, the descent speed might affect the proper retraction of the rope. You might need to slow down somewhat to allow the retraction motor to work properly.

I know of a local motor-glider tug that has a retractable rope system, where the light indicating the rope is retracted is operated by the winch motor back pressure. The problem here is that if there is back pressure on the motor due to something other than the rope being fully retracted, the light can come on with the rope fully or partly extended. It is however, often possible to verify the rope is fully retracted using the tug's mirror.

Some retractable rope systems also incorporate a regular hook for attaching a tow rope. There can be a swap-over mechanism that enables either the rope release or the guillotine. Before towing, it's obviously important to have this mechanism set correctly to carry out the correct function.

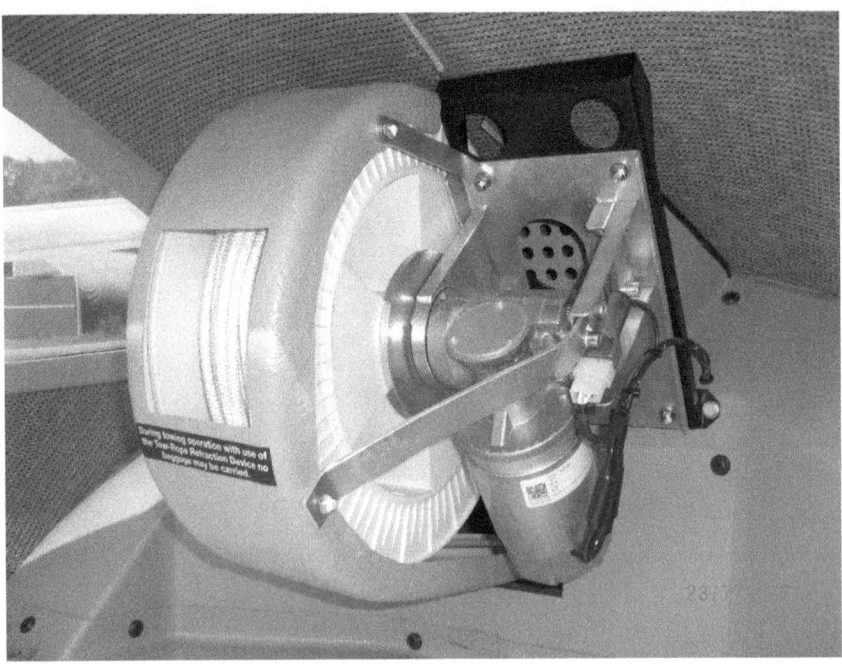

As part of the aircraft's daily check, the guillotine should be carefully inspected for damage, dirt, and debris.

Follow the manufacturer's servicing guidelines of course, but consider periodically:

- The guillotine could be tested to cut rope – maybe as part of scheduled maintenance – then sharpened and serviced as required
- The guillotine could be functionally checked, i.e. fired to check the mechanism at regular intervals. This is good opportunity for a more thorough check over and lubrication
- Some guillotines will damage themselves if they are fired when there's nothing to absorb the energy (i.e. rope) so it may be wise to insert a block of wood or something similar to absorb the impact (something that won't damage the blades but will absorb the energy)

The above scheme allows a good compromise between testing too often, which blunts the blades, and not testing enough – a mechanism gunged up with associated energy loss doesn't cut ropes.

This process also means pilots and engineers would be familiar with firing forces and the resetting procedure.

Guillotines are obviously extremely dangerous, so take all safety precautions!

Operational Considerations

It's the tug pilot's responsibility, after full consultation with the instructor in charge, to terminate aero-tow operations when poor weather or darkness approaches. Legal night-time (in the UK anyway) is defined as thirty minutes after sunset to thirty minutes before sunrise, determined by observation on the ground. Remember that near sunset time, the higher you get the brighter it is, therefore the lower you get the darker it becomes. Resist all forms of persuasion to launch when it's getting dark. Advise the instructor or person in charge of this time a little in advance if necessary. If you are operating outside legal daylight hours, you might invalidate your insurance.

It is also the tug pilot's responsibility to terminate aero-tow operations if weather conditions are deteriorating or the operation is becoming hazardous for any reason. Take personal responsibility for your operation and remember that, until release, you are the captain for the whole combination, no matter how senior the pilot of the glider may be.

Winch launching is popular in Europe because it is cheap. However, cables in the air or on the ground present a particular danger to operations. Before every take-off it is essential to check that a winch launch is not taking place or about to take place. A radio call is probably the best way of confirming this with the duty instructor or launch-point supervisor. At my club, we have an amber rotating light at the launch point that illuminates when a winch launch is in progress. Aerotowing should not take place close to winch cables lying out on the airfield because you might inadvertently pick it up with a wheel or skid.

Recording of towing operations is normally a legal requirement and is of vital importance to the club treasurer. Before take-off, note on the towing log card the glider registration and the name of pilot to be billed for the launch, and remember to set your altimeter to zero on the ground (QFE). Subsequently record the *exact* aero-tow release height (normally to the nearest 100 ft or 50 m). These records are essential for rendering launch charges to club members and must be legible and accurate. For example, a release at 2,100 feet should be recorded as such and not as 2,000 feet.

Passenger carrying is not normally good practice whilst aerotowing because it impairs the take-off and climb performance and adds an additional risk due to the potential hazards of aerotowing. Exceptions to this rule should have specific authorization from the club chief flying and gliding instructor, chief tug pilot, or deputy, and one should normally expect permission for instructional or operational reasons only.

Flights that involve landing away or flights for navigational practice would normally require logging out after first checking that the aircraft will not be needed during the time it's away.

In order to be safely conspicuous, anti-collision lights and landing lights, where fitted, should be on whenever the aircraft is in flight and for ground operations. Acknowledging that life of such lights can be limiting, it should be noted that there are some tug modifications available around the world that increase the life of the landing/taxi lamps. Be aware that high-intensity strobe lights should be turned off near refuelling installations as they pose an ignition risk.

Of course, no tug aircraft should take off unless the pilot in command is certain there is sufficient fuel in the tanks for safe operation. If in doubt, refuel! Fuel starvation is still the most common cause of engine failures in piston single-engine aircraft. It's imperative that you do not allow yourself to be pressured into carrying out 'just one more tow' when you think you should be refuelling. Many fuel gauges are inaccurate, and in some aircraft it is difficult or impossible to verify the fuel quantity visually. If in doubt, refuel to full, performance limitations permitting of course.

During towing operations, the electric fuel pump (if fitted) should be on when the engine is running to provide two fuel-supply sources and a positive fuel pressure.

There are advantages and disadvantages to turning the fuel cock off at the end of flying when the tug is put into the hangar. If the fuel is turned off, sooner or later someone will try to take off with it in that position. Risk assessment suggests it is best to simply leave it on, but consider shutting down the engine by occasionally selecting the fuel off. This exercises the off-cock and verifies it functions correctly. You might also be surprised at how long the engine will run with fuel selected off. Don't forget to turn it back on.

Towing Aircraft Allocation

A status board displaying allocation of tug aircraft is sometimes used by clubs. This could indicate which tug should be used first for maintenance or performance reasons. For economic reasons you might be better off allocating a four-cylinder towing aircraft rather than a six-cylinder aircraft. However, if performance issues are paramount, obviously the higher powered six-cylinder tug might be preferable – assuming your club is lucky enough to have the choice.

Engineering, Daily Inspection and Defects

Certified tug-type aircraft will normally have a 'supplemental' or 'change sheet' added to the operational notes for the purposes of the issue of the certificate of airworthiness for towing gliders. Here are some examples of additional limitations added to a Piper Cub (PA 18) tug type registered in the UK:

- The number of gliders on tow shall not exceed two
- The breaking load of the towing cable(s), or weak link(s) if fitted, shall not exceed 1,000 lbs (454 kg) when towing one glider. A weak link, the breaking load of which shall not exceed 1,300 lbs (590 kg), shall be installed between the towing bridle and the tug aircraft when towing two gliders
- Air speed shall not exceed the lesser of the maximum permitted speeds for the glider or gliders under tow
- A serviceable cylinder head temperature indicating system shall be installed. The cylinder head temperature shall not exceed 260 degrees centigrade (500 degrees Fahrenheit) and this limitation shall be marked with a red radial line on the cylinder head temperature indicator
- Outside ambient air temperature shall not exceed ISA + 15 degrees centigrade (ISA standard temperature is 15 degrees centigrade at sea level, so that would put a maximum outside air temperature of 30 degrees centigrade for this type)
- Towing procedures should be in accordance with those recommended in the British Gliding Association document entitled 'Notes for Tug Pilots' (which incidentally, is superseded by this more comprehensive book)

Your tug type should have similar limitations.

Any defects must be recorded and brought to the attention of the chief tug pilot or engineer immediately so that they can be rectified at the earliest opportunity and the tug prevented from operating in a dangerous state. In most states, it's normally a legal requirement to report aircraft defects.

Fuel and Oil

Pay particular attention to how much fuel you have.

- Don't trust your fuel gauges
- Don't trust your fuel gauges
- Don't trust your fuel gauges
- Lack of fuel accounts for a high proportion of engine failures

It is self evident that aircraft should not take-off with the low fuel light (if fitted) illuminated, especially as the nose-up attitude, in extremis, could lead to fuel surging in the tank away from the fuel-feed pipe. The Robin DR400 type advises that the last two gallons are not available in a high nose attitude such as take-off, climb, or go-around. To work out how much fuel you consume it is imperative you consult your particular type's pilot's notes but be aware that glider towing operations will mean that you will be operating under high engine power settings for much of the time. Estimate how many tows to a certain height you can safely accomplish with a full tank of fuel and leaving safe reserves.

Try to anticipate tow demand. Refuel when demand is slack and ready the tug with full fuel before peak demand occurs. For example, refuel before gets soarable.

Tugs should normally be refuelled at the end of the day unless a request has been made by the chief tug pilot or engineer for servicing purposes. A full tank of fuel is less susceptible to the formation of water by condensation and is also good for bladder tanks as it helps stop them from cracking.

Oil should be kept in a remote fuel store – ensure that the correct oil is being used before topping up the engine level and try to avoid overfilling.

Taxiing

When taxiing, bear in mind you might have a long rope behind you. Allow plenty of space to manoeuvre so that it doesn't snag on things or people. To help protect the undercarriage and reduce 'nodding', taxi from hard surface to soft surface and vice versa at an acute angle. When taxiing over damp grass when the humidity is high, you risk carburettor icing. Carburettor hot air is normally unfiltered, but as a conscious risk assessment, selecting it on during taxiing in these situations might be appropriate.

Hold the flying controls in the correct position with the stick normally held back but see the additional information section on taxiing in strong winds.

Ready to Tow?
(Before Flight)

Ensure the aircraft is fit for flight. The aircraft might have been checked at the beginning of the day's proceedings, but there are still many gotchas. So, check for things like refuel caps refitted, oil-replenishment caps refitted, cowlings secured properly, loose articles not left in the cockpit, tow rope incorrectly fitted, and tow-bars not still connected. I suggest at least a wide, 360-degree walk-round of the aircraft before getting in.

I once taxied out on an instructional flight with the tow-bar still connected. Fortunately it fell off on the airfield but potentially things could have been much worse, as it could have bounced up and hit the propeller or even fallen off in flight. I had assumed my capable student had prepared the aircraft properly and approached from behind, but moral of the story – never assume, check.

Is It Going to Perform OK?
(The Engine Start and Run-Up Check)

Before starting the engine, turn the tug so as to avoid streaming your prop wash into an open hangar, onto nearby parked gliders, or aircraft. On taildragger types it's important to hold control stick fully back to prevent propeller wash lifting the tail. Call 'clear prop' and *pause a moment before start* to enable anyone close to the propeller to move clear. Handle your particular engine in accordance with the pilot's operating notes. To clear any ice, the carburettor hot-air check (if fitted) should be done before the magneto check. Leave the carburettor hot air on long enough to clear any ice accumulation, a minimum of 15 seconds is recommended by the UK Civil Aviation Authority. After the power check and if there is any doubt whether the engine is developing full power, do not attempt to take-off until the problem has been rectified. A slight reduction in static (aircraft not moving) RPM equals a substantial reduction in power available. For example, an engine rated at 180 hp at 2,700 RPM will only produce about 150 to 160 hp if you try to take off at 2,400 RPM.

Before Flight Vital Actions

These checks help to ensure a successful take-off and flight. Normally, do at least what is suggested in the pilot's notes. Also refer to the section on engine management. If the aircraft is so fitted, I would also encourage the inclusion of turning the transponder on and selecting altitude encoding if available. This will enable a radar air-traffic controller to see you and your altitude, giving you and others extra protection. If you accidentally enter controlled airspace and have a serviceable transponder that you haven't turned on, the authorities are more likely to prosecute you. If your infringement is an honest mistake and the kit is turned on, in my opinion they are more likely to just give you a telling off. The other advantage of having the transponder on mode C (altitude encoder on) is that a TCAS equipped aircraft will get a warning and be able to avoid you. Consider the headlines after a mid-air collision:

Pilot had equipment to prevent this tragedy but didn't turn it on.

Although you probably won't be bothered!

Signalling

Radio communications are recommended, if possible. In the UK, we use the terms 'Take up slack' and 'All out'. 'Take up slack' is three words to differentiate against 'all out' which is, of course, two words – so don't say 'all out please', as this goes back to three words, which could cause confusion if misheard. Also don't say 'up slack', instead of 'take up slack' for the same reason. 'Stop' speaks for itself.

The waving of arms, bats, and other things is often used which has advantages and disadvantages. At some clubs, taking up slack and all out is done using the tug's mirror. On mainland Europe, sometimes the glider wing is left down, and then raised to indicate all out, and sometimes the glider airbrakes are used. Those are all fine, as long as everyone knows the system.

I consider it important to have a method of stopping a launch should something go wrong. In the UK, we've had a spate of gliders being unable to release at the top of the launch. In some cases, this has been caused by a snatch where the glider has slightly overrun, and the rope (or winch cable) became entangled in the main wheel or nose wheel. The British Gliding Association recommends stopping the launch in such cases. Anyway, there are many ways to skin a cat and the chosen best way is left for the individual clubs to decide.

Two recent incidents in the UK have been prevented from becoming tragedies because radio was used to stop a launch after it had started. In one case, a Ventus started its take-off roll with the tailplane not fastened at its leading edge. The launch point noticed and radioed the tug to stop the launch, therefore preventing the glider pilot taking off with a lethal mis-rigging of his glider.

The second was when a Robin DR400 tug caught fire as it took off, and the pilot was unaware of the problem. The launch point spotted the fire and told the tug pilot to land immediately as they were on fire! Both the tug and glider landed safely. The tug pilot quickly got out and ran away, and the aircraft was then totally destroyed by the fire! So, two cases of radio preventing potentially fatal accidents shows that safety can be enhanced by the use of radio for aerotow signalling and control.

Dave Bullock (BGA Glider Accident Investigator)

Take-off

Remember to expect an engine failure on every take-off. Plan for this on every tow; then, if it ever happens, you will be well prepared for a successful outcome.

Here is a report from the UK's *General Aviation Safety Information Leaflet* (GASIL), issue number 9 (October 2010).

Maximum power?

In a report from the BFU (German AAIB) we read of an accident to an aerotow combination just after take-off. The PA25 tug collided with trees, and the crew of the Janus Ce glider were seriously injured when it hit trees while manoeuvring to avoid collision with a house wall. The investigation was unable to find any apparent fault with the aeroplane, but the report notes that the carburettor hot-air selector was found in the fully out position.

We frequently remind pilots to be familiar with the contents of SafetySense leaflet 14 'Piston Engine Icing' and to warm the carburettor with hot air, if fitted, before take-off and at regular intervals. However, it is also vital that the hot air system is de-selected whenever maximum power is required from the engine. If a pilot suspects that his engine is not giving maximum power when asked, it should be automatic to check the position of the hot air control as well as that of the power lever and flaps.

Report reproduced by the UK CAA in General Aviation Safety Leaflet issue 9 (2010)

Move into position by passing across the nose of the glider to be towed in order to bring the rope closer to the ground crew. Park at about 90 degrees to the glider. This gives you a good view of the glider, the approach, and take-off area, whilst sending the slipstream clear. Try to take any wind into account so that the aircraft sits as much into wind as possible. This aids engine cooling and reduces the chance of controls banging against their stops while waiting for the next launch. Also consider parking so that people approaching the aircraft will be clear of the propeller. However, if you have a retractable rope on your aircraft you will have to park directly in front of and in line with the glider to ensure a clean pull out of the rope.

If waiting for extended periods, consider shutting the engine down, but be aware of a quick start followed by a quick take-off with a cold engine or forgetting things. It's not normally necessary to complete an engine run-up before every tow as you have just flown the aircraft – but please do one if you suspect any engine problems. *It is necessary to complete the pre-take-off vital actions appropriate to the type of machine you are towing with.*

As an example, the minimum checks considered 'vitals' for a typical fixed-pitch prop tug, assuming you haven't shut down, might include:

- Resetting the elevator trimmer
- Checking there is sufficient fuel for the next tow
- Setting the flaps for take-off
- Checking the engine temperatures and pressures (Ts&Ps) are correct
- Ensuring hatches are secure
- Ensuring carburettor hot air is off (cold)

I also think a magneto 'drop not stop' check is advisable after each start and occasionally during towing operations. I have learned of a tug airplane flying a number of glider launches on only one serviceable magneto. The failed magneto should be picked up when full power is applied for take-off as the RPM or manifold pressure will be lower than normal. However, I know of a number of experienced pilots who have missed this, so my advice is to check each magneto before take-off, if there is any doubt.

If fitted, the fuel pump should normally be left on throughout towing operations and the mixture normally at rich to aid cylinder cooling. Some pilots lean the mixture in the climb to save fuel, but in my experience most pilots don't. Leaning the mixture can cause problems if not done properly; and it takes time to do properly, which will potentially diminish pilot lookout.

In my club, a little checklist is printed at the top of the normal checklist and can be read as it sticks out of the stowage pocket. Consider adapting the 'before take-off' checklist to a transit checklist appropriate to your machine. Use the pilot's operating notes as a basis.
Before launching:

- Notice the call sign of the glider you are about to tow – consider a radio check with that glider
- Watch out for rough ground in the take-off area
- Be very aware of cables laid out on the field
- Beware of winch launch tow hooks on the glider – they should not be used for aerotowing if there is a nose hook installed
- Beware of low-experience glider pilots
- Beware of short tow ropes
- Beware of light-weight glider pilots that might cause the glider they are flying to have a rearward centre of gravity
- Beware of turbulence and wind shear
- Beware of cross or slight tail winds
- Be alert to obstructions out to one side at an angle that could introduce an obstruction in a swing or affect your climb-out or engine failure options
- Check that glider tail-dolly is removed – this is primarily the glider pilot's responsibility but a rearward centre of gravity could result in a tug upset

A general rule of thumb is that if you have any three of these things against you, do not launch.

Once the ground crew are clear and you get the take up slack signal, move into line and take up slack by moving gently forward using minimum power and without brakes. Riding the brakes not only increases wear but warms them up, which renders them less effective when you need them. It's important to position the tug in line with and pointing in the same direction as the glider. Depending on wind direction, surface, and glider type, there is a good chance that if the combination isn't lined up properly to start with, that the glider will either drop a wing shortly after it starts moving or will run straight and uncontrollably in whichever direction it happens to be pointing.

Scan ahead, to the sides, and include looking in the mirror. If you are not happy that the take-off run is clear, don't launch. At the 'all out' signal, take two seconds or more on a powerful tug to move the throttle to fully open. As you reach full power, it is vital to check that full power is being delivered. This is the only opportunity you have to check for this, and the tow should be rejected (your first action is normally to release the glider – see emergency procedures) if the RPM or manifold pressure are not within the range indicated by the aircraft notes. I have marked an indelible reference mark on the tachometer of the tow aircraft I normally fly at the normal static (not moving) RPM. If the engine is not achieving full static RPM it might be something simple like the carburettor hot air being left on or one magneto being off.

Meanwhile, it is important to keep straight on the take-off run as some gliders, particularly those with nose-wheels, have less steering ability than the tug. It's normal to keep your hand on the throttle to prevent a slight creep from full power and subsequent large reduction in the take-off power produced. It's also essential to have the rope release near to that hand in the event of having to release the glider. Pre-planned or rehearsed abandonment drills have been proven to benefit the successful outcome in an undesirable event.

It's probably worth considering the glider pilot's problems induced by a powerful tug's propeller wash combined with a relatively short tow rope. This propeller wash can contribute to a glider dropping a wing and subsequent ground-loop, which in turn of course will affect you in the tug. This is hypothetical and as far as I know, hasn't been proven, but many glider pilots will have noticed a wing drop without knowing what could have caused it:

- In nil wind or headwind conditions the glider pilot could drop the wing associated with the down-going propeller blade due to the corkscrew effect of the tug's propeller wash streaming down the glider's fuselage, lifting one wing and lowering the other

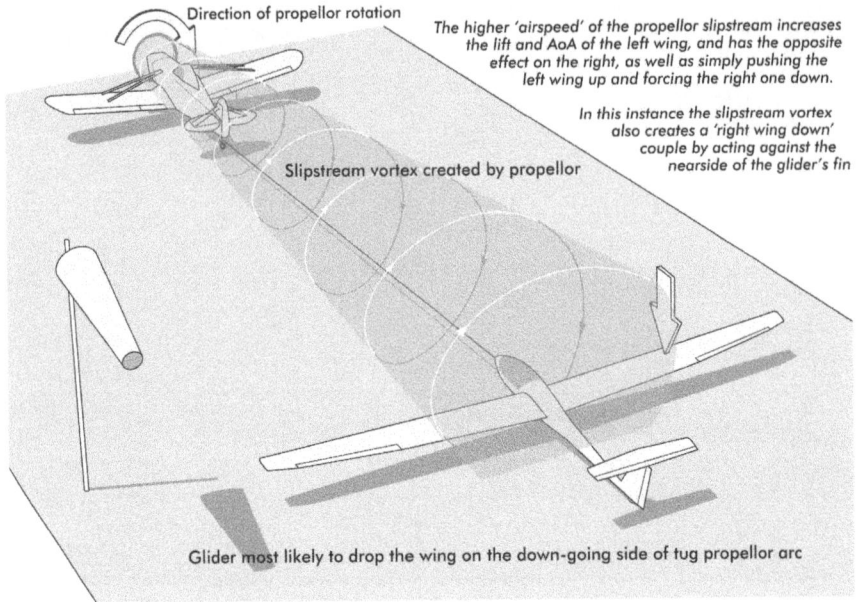

Direction of propeller rotation

The higher 'airspeed' of the propellor slipstream increases the lift and AoA of the left wing, and has the opposite effect on the right, as well as simply pushing the left wing up and forcing the right one down.

In this instance the slipstream vortex also creates a 'right wing down' couple by acting against the nearside of the glider's fin

Slipstream vortex created by propellor

Glider most likely to drop the wing on the down-going side of tug propellor arc

- In cross-wind conditions the glider pilot could drop the upwind wing as the tug's propeller wash streams over the glider's down-wind wing, invigorating lift over that wing and causing it to rise and of course the other wing to drop

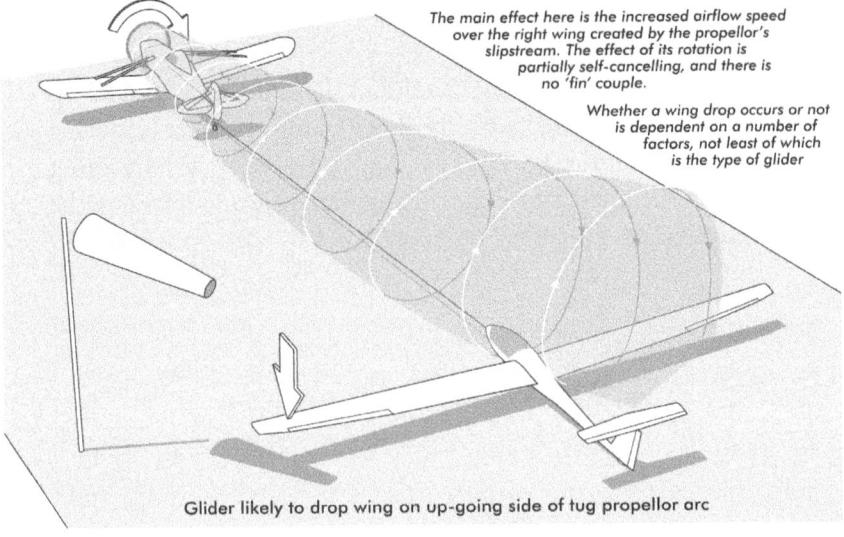

The main effect here is the increased airflow speed over the right wing created by the propellor's slipstream. The effect of its rotation is partially self-cancelling, and there is no 'fin' couple.

Whether a wing drop occurs or not is dependent on a number of factors, not least of which is the type of glider

Glider likely to drop wing on up-going side of tug propellor arc

Nothing concentrates the mind like reading about and learning from an accident, or a near accident. i.e. trusting an instructor on my wingtip to have his brain in gear, I commenced towing a Libelle without looking at the winch launch point myself, and they launched just as I took off. Because the Libelle was light it towed straight out rather than moving toward the local valley. The first I knew there had been a winch launch was noticing the parachute descending about 20 feet from my wingtip … … … .

The most scared I have ever been. Came down, apologized to all, and made a rule. Tuggie ALWAYS KEEPS AN EYE ON WHAT IS GOING ON AT THE WINCH LAUNCHPOINT! Trust nobody! Not even me. (That's another story!)

Mary Meagher

Performance

Consider the performance of the combined tug aircraft and glider. Remember that water-ballasted gliders, long grass, wet grass, upslope, high altitude, high temperature light wind, or tail wind, will all increase your take-off run. Obstacles such as tall trees at the end of the airfield and in your take-off path might pose a problem. It's worth considering a decision point or a point where if you are not airborne you might reject the take-off, starting with releasing the glider and stopping ahead or getting airborne without the glider. Some pilots advocate reaching two thirds of your take-off speed by one third of the available take-off distance. It's impossible to have performance calculations for every combination of tug and glider. For example, you could be towing a single seater, a two seater, a glider with or without water ballast, or a glider with a nose skid that takes a few painful yards to get off the ground. Towing off a hard surface will normally take less of a ground run than towing off grass. If towing off grass or soft ground, consider your aircraft's soft-field technique.
In all cases, apply lots of common sense!

By UK definition, 'take-off run' is that required to getting airborne and 'take-off distance' is that required to fifty feet height.
The UK CAA Safety Sense Leaflet number 7, 'Aeroplane Performance' suggests increasing take off distance to height 50 ft by:

- 10 per cent for every 1,000 feet of density altitude
- 10 per cent for every 10 degrees C increase in temperature above International Standard Atmosphere (ISA)
- 20 per cent for up to 8 inches (20 cm) of dry grass
- 30 per cent for up to 8 inches (20 cm) of wet grass
- 10 per cent for 2 per cent uphill slope (add no advantage for down slope)
- 20 per cent for tailwind component of 10 per cent of lift-off speed
- 25 per cent or more for snow or soft ground

Then, also add another safety factor of 1.33.

Take into account your lack of practice, possible incorrect speeds and techniques, aeroplane/engine wear and tear, and less than favourable conditions.

Low tyre pressure (perhaps hidden by grass or wheel fairings) will increase the take- off run, as will wheel fairings jammed full of mud, grass, slush, etc.

Rain drops, mud, insects, and ice have a significant effect on aeroplanes, particularly those with laminar-flow aerofoils.

Slopes can be calculated if surface elevation information is available. If not, they could be estimated. For example, a height difference of 50 ft on a 2,500 ft (750 m) strip indicates a 2 per cent slope.

On some types, a cross-wind on take-off may require use of the brakes to keep straight and will increase the take-off distance.

If towing from an unknown length, consider pacing it out. The pace length should be established accurately or assumed to be no more than 0.75 metres (2.5 ft). It might be better to measure the length accurately with the aid of a rope of known length or even an accurate GPS receiver.

One summer evening, long ago, a few of us were in the hangar looking out to the west at some obvious wave lenticulars. One very experienced pilot decided it might be worth a try at Diamond Height so pushed out an oxygen equipped Astir. I was elected from a cast of one to fly our Gypsy engined Chipmunk. The windsock was hanging slack, as we lined up heading west. The take off seemed even more prolonged than is normal following a Gypsy, but there was about 1000 metres in front of us. Eventually the combination staggered airborne. As I passed the windsock on the western boundary, I noticed at least 10kts of easterly wind had appeared. Immediately after the windsock runs a motorway, which was crossed very quickly and between passing high sided vehicles. Four fields after the motorway, our previously impressive groundspeed suddenly dropped to next to nothing and the ground started to look less close. The Astir eventually released in good lift at 1800 feet. When I got back to the airfield, the windsock was showing 25kt westerly!

I learned about rotor from that! Oh, yes, he did get his Diamond Height.

Paul Whitehead

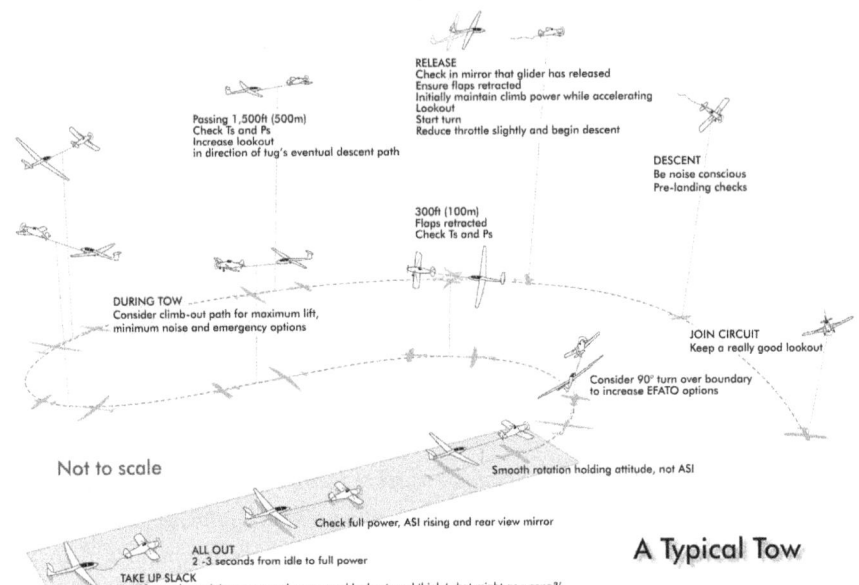

RELEASE
Check in mirror that glider has released
Ensure flaps retracted
Initially maintain climb power while accelerating
Lookout
Start turn
Reduce throttle slightly and begin descent

Passing 1,500ft (500m)
Check Ts and Ps
Increase lookout
in direction of tug's eventual descent path

DESCENT
Be noise conscious
Pre-landing checks

300ft (100m)
Flaps retracted
Check Ts and Ps

DURING TOW
Consider climb-out path for maximum lift,
minimum noise and emergency options

JOIN CIRCUIT
Keep a really good lookout

Consider 90° turn over boundary
to increase EFATO options

Not to scale

Smooth rotation holding attitude, not ASI

Check full power, ASI rising and rear view mirror

ALL OUT
2 -3 seconds from idle to full power
TAKE UP SLACK
Inch forward on minimum power, keep a good lookout, and think 'what might go wrong?'

A Typical Tow

Airborne

Once airborne, smoothly make the transition into the climb, allowing the tug to accelerate. If you are towing a heavy glider, it might be beneficial to unstick and accelerate in ground effect. Try to avoid any sudden pull-up as the glider pilot might also aggressively pull-up in an attempt to maintain station behind the tug but overdo it and overshoot into a tug upset situation (described further in emergency procedures). Any pitch or speed changes should be very smooth and gradual and made by primary reference to the attitude. Remember that tugs can normally manoeuvre more easily than gliders. If there is a wind gradient, accept a slight increase in speed as you climb rather that zooming up into the gradient leaving the glider behind in the lower wind with less energy. (See section on strong winds) Heavily ballasted gliders normally require extra speed soon after take-off and hence during the tow. It's interesting that some heavy gliders appear to fly around quite happily at say 50 knots (93 kph) but struggle when towed at a similar speed. There's been some hot debate about this, but I still don't really understand why.

Slow and Fast Gliders

Certain vintage gliders have very slow max aerotow speeds. The minimum safe towing speed could be around 55 knots (100 kph), depending on type of course, and should not be much lower as the tug's engine could overheat, and low-speed control problems could result. Consider using part flap for the tow if necessary, but remember to retract it after release and before accelerating to prevent potential over-speed of the flaps. Monitor the CHT (Cylinder Head Temperature) carefully, and weave frequently as the higher nose attitude might obscure your view, although some flap usage could help alleviate this by lowering the nose. If turning tightly in thermals at low speed, consider the inevitable increase in stalling speed due to increased wing loading. Stalling speed will increase by an amount proportional to the square-root of the wing loading (or G loading); therefore, stalling speed will increase with increasing bank angle.

High-performance aerobatic gliders usually require towing at a higher than normal speed – check with the aerobatic pilot. (See section of aerobatic towing)

It will usually be worth considering flying at faster speeds when towing cross-country, but remember to check the glider's maximum aerotow speed.

Full Climb

The route should normally be planned to fly under areas of lift whilst avoiding the sink. Turns should be shallow, but with experienced glider pilots more bank angle could be utilized to maximize lift in thermals.

The first turn should normally be into wind to keep the glider within easy gliding range of the field at all times. When low, stay over areas where the tug could successfully land after an engine failure. If the tug has a safe option, the glider pilot should have one also. Avoid turning through or flying directly into the glare of the sun if possible, especially in winter when the sun is low. As previously mentioned in slow flight above, your stalling speed will increase in a turn – quite a lot in a tight turn. If turning tightly in thermals at low speed, consider the inevitable increase in stalling speed due to increased wing loading. As previously mentioned, stalling speed will increase by an amount proportional to the square root of the wing loading. If the wing loading is increased by four (four Gs), the stalling speed will double, because the square root of four is two.

During the climb, try to vary your heading at intervals to reduce the risk of collision. This aids good lookout. Manoeuvring also shows you up to other aviators – it's like shouting 'I'm over here!'

Be aware that a club two-seater will usually be training, so remain close to the field and avoid excessive manoeuvring. For spinning exercises, the glider will need to be fairly close to the airfield. A cross-country pilot, in contrast, will probably want a fast, direct tow to the nearest source of lift and will normally be happy to be released further from the field. Don't tow gliders to the down-wind side of the site unless you have a good reason. Always bear in mind that the glider may release at any time and should still have had a beneficial launch. Consider, a student will not be happy paying for a 2,000 feet launch from which he has to spend all the flight time flying directly back to the airfield instead of practicing manoeuvres.

Keeping the Neighbours Happy
(Noise-Abatement Procedures)

All tug pilots must be aware of the possibility of noise complaints and avoid towns, villages, and farms by the widest practical margins. Consider passing down-wind so as that 'drifting noise' is blown away from such places.

Continual towing or descent over the same area may cause considerable nuisance and irritation to your neighbours. Some tugs might have already been modified to minimize the actual noise they produce, but we can also spread the load by thoughtful and varied tow-out patterns. It is variation that should form the basis for our noise-abatement procedures.

The following general points should be considered:

- Make full use of all airspace available to you
- It is not always necessary to drop upwind; a tow made for the most part down-wind of the site and then terminate overhead or slightly upwind of the site should also be considered in certain situations but try to keep the glider within gliding range of the field
- Remember that when turning, the focal point of your turn (the lower wing will be pointing at it) will be subjected to a concentration of tug noise, assuming no wind
- A soaring pilot may be happy to be towed directly away from the site; this should be done when the opportunity arises and the glider pilot agrees!
- Noise carries down-wind, therefore consider passing down-wind of noise sensitive areas
- Manoeuvres can make you more conspicuous and aid a good lookout

Remember that the noise of a descending tug with a relatively high engine-power setting can be equally annoying, so apply similar techniques in descent. Try and make your descent route different from the tow-out route, and don't descend too low before entering the circuit area to assist in reducing noise to your nearest neighbours. Be very careful not to descend onto circuit traffic that might be in your blind-spot.

Think about circuit heights that could be lower than glider circuits for low-wing aircraft such as the Pawnee or Robin and higher than glider circuits for high-wing aircraft such as the Super Cub, to facilitate a good lookout.

Consider planning and flying a continuous descent (or energy) profile, which could save fuel and reduce noise.

Glider in Low-Tow Position

If the tug requires sustained forward stick pressure, the glider has possibly gone into the low-tow position. In this case, the glider will not normally be visible in the mirror, but the rate of climb and forward stick pressure indicate that it has not released. Continue to use normal towing speeds. The glider should return to the normal tow before release, otherwise it might not be obvious to the tug pilot when the glider pilot has released. There is also an added chance of the rope rings coming up and damaging the glider or canopy. Normally the glider cannot get too low on tow, but if the tug's stick reaches the forward stop and the airspeed begins to decay, wave-off or release the glider. Don't immediately begin to descend as you may then come into conflict with the released glider.

Some pilots consider the low-tow position to be more stable. It is sometimes preferred for cross-country retrieves as it has the effect of towing the glider 'uphill'. Some might argue that the other advantage of the low-tow position could include less chance of a tug upset. But then again, if the glider pilot zooms up from an even lower starting position, it could increase the chances of a 'slingshot', which could make matters worse (See emergencies). Some pilots suggest that low tow is not appropriate in level flight and in turbulence, because if a longish rope surges, it can induce a pitch up in both glider and tug. Surging appears to be less marked for tows in the normal position, so the effect is less evident. Some countries do low tows as the norm.

Where's the Glider Gone?
(Glider out of Position)

Glider pilots often practice 'boxing the slipstream'. For the low tow, continue to fly the tug's attitude as described above. Firm, sustained rudder inputs will be needed if the glider is out to one side, but be cautious of using too much rudder (See emergency procedures). Try not to turn toward the glider if it is out to one side, as the resulting bow could snatch hard enough to break the weak-link or rope. If the glider pilot does cause control problems, or demonstrates erratic flying that may put the tug at risk, wave the glider off, or if necessary, release it from the tug.

Glider Release

First and foremost *keep a good look-out!* Glider release can normally be felt but confirm release. Utilize the mirror if necessary to see that the glider has really gone and note the direction of release. If in doubt, consider a positive wing-waggle before descending, this should indicate to a glider pilot still on tow to release immediately. Firstly, check that the flaps are retracted. You might have been towing a low-speed glider and need to check that you won't accelerate out of the airspeed indicator white arc speed and overstress the flaps. Normally fly the tug ahead on full power, RPM/power limits permitting, for a few moments to ensure maximum separation and aid engine cooling. Beginning a slight descent would promote an increase in speed and also increase separation. A turn in the opposite direction to the glider would increase separation between the rope rings and the glider. The glider pilot would normally pull up and slow down, but could turn either way. Please note that there are different rules around the world about which way tugs and gliders should turn after release. In some places on mainland Europe, the tug will turn left and the glider pilot will turn right. The military club I first learned at had the exact opposite rule. Also consider your actions in special situations, such as dropping a glider on a ridge, where it might be unwise for either the tug or the glider to turn towards the ridge.

You might have a tug with a retractable winch system. Don't forget to retract the rope for landing. (See previous section on retractable ropes).

I was towing in a high wing aircraft and in accordance with good practice to clear the turn I banked slightly the other way, raising the wing to have a good look in the direction I wanted to turn. Shortly after this turn the glider pilot released. I subsequently found out that he thought I was waving him off!

Anonymous

Descent

For more specific information, see the separate section on engine handling and your aircraft's specific notes.

After glider release, the engine is extremely hot, and precise engine handling is required to prevent shock cooling damage, otherwise known as thermal stress. The cylinder head and the valve seats cool at different rates, so too low a power setting for the descent will cool the cylinders whilst the valve seat remains hot and expanded, causing a stress crack in the cylinder head. The front cylinders are the most vulnerable.

Take sufficient time to accelerate to the correct descent speed whilst retarding the throttle to maintain the climb RPM or power and paying particular attention to the engine limitations. When you become familiar with your specific type you might be able to adjust the descent power by throttle position and sound rather than looking at the engine instruments, which aids lookout. There should be no notable change in engine note at this stage. Once stabilized, take a further fifteen seconds or so to gently reduce to the descent RPM or power and maintain the correct descent speed, trimming as required. Continuous use of carburettor hot air will be required at low-power settings. However, at the higher-power settings associated with a tug descent, it might not be appropriate. Fifteen seconds of carburettor hot air applied during the descent on a moderate power setting should normally be enough to melt any accumulated ice. Carburettor hot air should not be applied at all during full-power settings as this causes detonation, reduces available climb power, and can damage the engine. It *is* possible to get carburettor icing with full power. If this is suspected, reduce power and then apply carburettor hot air long enough to melt the ice.

Continue to avoid noise sensitive areas during the descent and try to fly in the sink between thermals. Turning will improve your rate of descent and make you more conspicuous to other flyers. Sometimes side-slipping will increase your rate of descent, make you more conspicuous, aid lookout, and depending on type, might reduce the chances of engine-shock cooling. Consider the manoeuvring speed (Va). A good lookout is imperative as the tug is now travelling much faster than most gliders nearing the circuit. At winching sites, try to avoid the vicinity of the winch run, as it may not be obvious from the ground if you are clear of the launch path or not.

I was converting a PPL to our tug (a Lycoming powered Chipmunk). The student (in the front cockpit) had just performed a poorly judged glide approach onto the runway. I decided to demonstrate another glide circuit and called for a go around. This was done, and I took control during the climb to rebrief. On reaching our high key point at 2000', I tried to throttle back. The lever moved, but the RPM did not! I asked the student to try his throttle lever, which he did, but to no effect. I lowered the nose to avoid entering cloud at 2200' which caused the RPM to rise to near max permitted, and increasing. Banking to about 70 degrees and pulling hard stopped the speed increase, and thus the RPM increase. We were now stable in a high 'G' turn, just below an overcast cloudbase, wondering what to do next. Selecting carb heat caused the usual small decrease in power, but not enough to even reduce the G significantly, never mind start a descent. Slowly we came to the conclusion that we would have to shut the engine down, although that might cause shock cooling. We did not think about turning just one magneto off, although I strongly suspect that would not have produced enough power reduction anyway. Eventually I got the front seat pilot to pull the mixture control to cut-off, and turn both magnetos off. That achieved a comprehensive RPM reduction, all the way to zero in fact, as we slowed down. The resulting PFL was uneventful, even serving as the demo I had promised to my student.

The cause of the stuck throttle turned out to be an internal break in the Bowden cable running from the pilot's throttle levers into the engine, allowing an increase in power, but not a decrease.

Paul Whitehead

Circuit and Approach

It is essential that tug aircraft do not descend onto the glider-circuit pattern, which increases potential collision risks because you are descending into your blind-spot. You could plan the descent to join the circuit level approximately midway down-wind or on base leg. This is similar to a standard join procedure in the USA. Joining the circuit on base leg or even finals may be appropriate, but keep a good lookout and adopt the rules of the air, which normally require powered aircraft to give way to gliders. Always try to think ahead. Holding level could allow the aircraft to decelerate into the white-arc, flap-limit speed without you touching the throttle. Be careful not to cut up or land directly in front of a glider. If you need more room to sort out your circuit and stabilize your approach – take it. Select flap as necessary and position for a steep approach, reducing the throttle steadily to idle before touchdown. This is to ensure maximum clearance of the rope in the later stages of the approach and provide for smooth engine handling in the transition from descent power to touchdown.

At some airfields it's standard procedure to drop the tow rope before landing. This should be done at the designated drop area and normally in the direction of take-off and landing.

If in Doubt, Throw It Away
(Go-Arounds)

Notice that I have added go-arounds as normal procedures, rather than as emergencies. I did so to indicate that there is nothing wrong with going around. It's essential that a missed approach is carried out if you are unhappy with the approach or landing area for any reason – sometimes it can be just a gut feeling. A go-around is, however, potentially hazardous because of the possibility of a winch launch in progress, a glider on an abnormal approach, snagging your tow rope on something or someone, or other gotchas, but it is normally better than proceeding with a dodgy approach and landing. Always try to think ahead and be prepared for a go-around. If the landing area is blocked, make an early decision, as some ropes can be 200 ft (60 m) long or more and trail far below the tug. Depending on circumstances, you could consider dropping the tow rope.

Normally, the procedure would be to apply full power while simultaneously rotating to the climbing attitude and retracting the drag-flap, but, of course, operate in accordance with the pilot's notes for the particular type you are flying. Keep a really good lookout. Consider turning to move away from any winch run or other operation without cutting across the normal circuit pattern, and when you are ready, rejoin the circuit. I emphasize: if there is any doubt about a safe landing, go-around. There should be no consequences associated with any decision to go around rather than continue an unsafe approach and landing. The only time a go-around might be a problem is if you're short of fuel – but you shouldn't be.

Be very careful about doing S-turns or orbits on final approach as it is too easy to miss a glider on final glide or another aircraft in the circuit.

And remember, if there is a risk of ground contact outside the airfield or with any objects with the rope, release the rope during the go around.

The Landing

In general, do not fly over people, aircraft, or vehicles prior to landing. Remember, you have a long trailing rope that will hang well below your aircraft, so allow for it. Don't land too deep or you are in danger of what is known in the industry as a 'runway excursion'. That means going off the runway end (or side)! If your landing run is performance limited, consider dropping the tow rope and then land without it, which will require less landing distance. As with most aviation stuff, there is of course a balance. Pick a clear area and expect the unexpected, for example a vehicle driving out in front of you.

Some tug pilots retract the flaps immediately after landing in an attempt to firmly keep the machine on the ground and to make the brakes more effective. All very well, but I know of some who might argue against it.

Immediately after landing and during the roll-out I decided to raise the flaps, unfortunately I put the landing gear up by mistake ...
                                                                    Anonymous

When at slow taxi speed, consider which way to vacate the runway or landing area, bearing in mind that approaching aircraft may be expecting you to turn left as that is normal air-law procedure. However, it's probably best practice to turn away from any winch-run operation. Ensure there is a standard that everyone understands. Taxi back, and if the next glider pilot is not completely ready to launch, consider shutting the engine down. Plan where you park or stop so as to minimize obstruction and any ground running time needed to re-position the aircraft for the next launch.

Shutdown

You would not normally stop the engine while still moving; this looks slick but isn't really necessary. If you have just landed, idle the engine for approximately one minute, which will allow engine temperatures to stabilize.

Switch all non-essential electrics off, including fuel pump and radio.

You would typically shutdown the engine by closing the throttle and then pulling the mixture control to Idle Cut-Off (ICO). The engine can be shut down quickly in an emergency by selecting the magnetos off (See emergency procedures).

After the engine stops, immediately switch the magnetos off, remove the key (if there is one), place it where it can be easily seen from outside, and set the master switch to off.

Periodically consider turning off the engine by selecting the fuel off. This will exercise the fuel cock which may not be used that often, and also verifies its correct functioning. You might be surprised how long an engine will run in this condition. *Please don't forget to turn the fuel back on!* Most operators leave the main fuel selector on as a conscious risk assessment – this reduces the chances of somebody taking-off with the fuel turned off, which is more probable in a towing environment. However, some clubs turn the fuel off after the last flight of the day and then turn it on again for the daily inspection, or 'check A', to reduce the risk of potential leaking fuel contributing to fire in the hangar.

On some aircraft, for example the PA25 Pawnee, the fuel cut off valve is operated by a push/pull cable and knob. This will turn the fuel off satisfactorily, but because you are 'pushing a cable', may not fully turn the fuel valve on again. In these cases, an engineer should be consulted to ensure that the fuel valve is fully 'on' before the aircraft is flown again.

Dave Bullock

Cleaning

Tug aircraft should always be washed after flying to remove mud and dirt and, importantly, to remove corrosive exhaust emissions. Don't spray cold water directly onto a hot engine, this can cause shock cooling damage. After removal of large quantities of mud, consider re-lubricating control hinges etc. Take care if you use a jet wash, as they can be very good at blasting lubricants away and at lifting paint that is not well bonded to the skin. Windscreens must be kept clean throughout the day, using a clean cloth and water, or non-silicon based spray polish and a soft cloth. Harsh tissues scratch Perspex, so avoid them. Remember to also clean the rear-view mirror.

As an accident investigator, I know of a fatal airliner accident, in which the application of a jet of water contributed to the loss of that aircraft by the water subsequently freezing the angle of attack sensor.

Putting to Bed
(Hangaring)

Tugs are often the last aircraft to go in the hanger and so are located in the front. The parking brakes are normally left off so the aircraft can be moved quickly in the event of fire. It's worth rechecking that the aircraft magnetos switches are off for safety and master switch off to prevent a flat battery next morning. Some pilots turn the fuel off, but as mentioned in these notes elsewhere, I think it is best left on. As sure as eggs are eggs, someone will sooner or later try to get airborne with it turned off. Ensure that there is a club philosophy that every pilot is aware of. Fit pitot covers, canopy covers, and full covers if you are lucky enough to have them. Many clubs have blanks that protect birds from nesting the warm engine inlet area. Make sure this is a big 'Remove before Flight' flag on important stuff that could be easily missed, such as the pilot/static covers, engine blanks, etc.

Getting Feedback from Pilots after an Aero-Tow

Many aero tows are successfully carried out with little or no communications. This would normally indicate, of course, that all was well with that tow. Occasionally, the glider pilot or other observers might have helpful comments – these should be conveyed to the tug pilot to help improve the operation or enhance safety. Sometimes, simple things like a request from the glider pilot for more speed can be made over the radio. The glider pilot should make a mental note of the tug's call sign before launching in case he needs to make contact by radio during the tow. I consider radio communications very important – not really for general operations and chit-chat, but in case of an emergency. For example, if someone on the ground sees something about to go dangerously wrong on a launch they can warn those concerned over the radio. Fortunately, good, portable, air-band radios are relatively inexpensive these days, but unfortunately they can easily go missing in a club environment.

Section 2 - Emergency Procedures

Aborting a Tow on the Ground

If there is a problem with the tug, the pilot should *immediately release the tow rope.* This has two functions:

- Firstly, it immediately increases the separation between tug and glider
- Secondly, and most importantly, once the glider pilot sees they are 'pushing the rope' they are likely to be fairly convinced that getting airborne is improbable and should then be prepared for the subsequent ground run and to avoid the tug.

The tug pilot should consider stopping ahead if there is sufficient space, but do so without heavy braking if possible and attempt to steer gently to one side if safe. At any point beyond this, be aware of the possibility of the glider rolling into the tug – the tug brakes are usually much more effective than the glider's. The danger point is while the tug is still firmly on the ground and a light glider lifts into ground effect, losing the ability to stop quickly.

If the *glider pilot* aborts the tow, it could be safer to climb away, leaving the runway clear for the glider, and continue into the circuit with a similar procedure as for a go around. If the *tug pilot* aborts the tow because of, say, lower than expected RPM, consider releasing the glider from the tug and continuing forward to reduce the chances of the glider colliding with the tug from behind, as described earlier. In that example of reduced RPM, getting airborne is probably not a good idea, even without the glider on tow.

If it is not possible for the tug to continue the take-off – this time, for example, let's say the engine has failed – then roll to a stop without hard braking and watch the mirror, moving to one side if it's appropriate.

There comes a problem if the launch is aborted due to third-party input, for example someone shouting 'stop!' over the glider-radio frequency. The problem here is that the observer may be able to see a problem but the pilots may not. It's probably best to first release the glider, but would it be best for the tug to get airborne? Probably not if it has just picked up a winch-launch cable or the tug is on fire! I know of a tug catching fire on take-off because of cut grass accumulations in the wheel spats.

I suffered a partial engine failure, just as the tug and glider were airborne without room to land either ahead. I continued the climb and turned down-wind of the field to wave off the glider, but at 250-300ft (100meters) it took two wave-offs and a radio call before the glider released. There was an element of panic and disbelief with the glider pilot.

Guy Westgate

Glider Airbrakes Open

Glider airbrakes may open in turbulence, because the pilot failed to lock them properly or because of technical malfunction. If this is noticed early in the take-off roll, before the tug is airborne, consider releasing the glider so that it does not become airborne, and then closing the throttle and rolling to the end of the runway. If the tug is climbing at a poor rate, check that the throttle is fully forward (you should normally have your hand on it anyway), carburettor heat is off , both magnetos are selected 'on', and the engine gauges are normal and then check the mirror. If the glider brakes are open, *do not signal immediately unless absolutely necessary* but try to get the glider pilot to a safe height if possible and into a position where, if the glider pilot releases, they can land safely back on the airfield or over landable fields that may be reached even with full airbrakes – *then signal*. Consider: a dazed glider pilot is likely to just release at *any* sign of a problem. *However, if the tug is at risk, wave the glider off or release without delay.*

The signal for glider airbrakes open in the many countries is to *waggle the rudder rapidly.* Try to make this an obvious signal to the glider pilot, or simply tell the glider pilot to close their airbrakes or release his drag chute over the radio. It is the rudder waggle that is the signal to the glider pilot, not the yawing of the tug, which is best avoided, especially at low speed. If the signal is not well done, the glider pilot can mistake the roll resulting from the secondary effect of yaw for a wave off, which could lead to an unnecessary release and possible accident. As always, consider using the radio to identify and discuss the problem but only at a safe height; don't let its use distract you from flying the aircraft.

Remember the old but relevant adage:

- Aviate
- Navigate
- Communicate
- *In that order!*

Some pilots start with the airbrakes open because they feel that the glider's aileron response is not good on the start of the take off roll, and that, with the airbrakes open, they 'spill the air outboard', therefore increasing the airflow/ effectiveness of the ailerons. On some gliders, this is true, but not very many. The ASW24 is one of the few gliders where it is usual, and probably justified.

A lot of pilots do this just because 'they think that it looks cool', and actually is potentially quite dangerous. Even an ASW24 has acceptable aileron response, except for days with no headwind, or light cross-wind, and a not very powerful tug. The problem with starting the take off roll with the airbrakes open is that, if the wing does drop, the pilot has not got his hand on the release, and is likely to be slow to release. I've heard of an ASW24 that was broken because, when the pilot was taking off in it with the airbrakes open, he dropped a wing and could not release in time, ground-looping the glider and causing substantial damage. Personally, I think that in the vast majority of occasions, taking off with the brakes open actually increases the risk of an accident because of the delay in getting to the release if required. The exception would be, as stated above, in an ASW24 or similar, no headwind and a low powered tug. Even in this case, I would start off with my hand on the release, even with the brakes open, and when aileron response was adequate, move my hand from the release to close the brakes, in much the same way as one would set negative flap on some flapped gliders, but still start off with your hand on the release, and only let go of the release to change the flap setting once aileron response was assured.

Dave Bullock

I would add to Dave's comments. If the glider pilot intends starting the take-off run with the airbrakes open to increase aileron control, *inform the tug pilot.* He might just dump you thinking you have forgotten to lock them.

Note: It is British Gliding Association Standard Operating Procedure for the glider pilots to have their hands on the release knob for the start of the take-off.

Tug Unable to Keep Straight on Take-off Roll

If the tug pilot is unable to keep straight on the take-off roll, it might be because the glider pilot has dropped a wing and is pulling the tug's tail round. Check your mirror or look over your shoulder (if you can). If this is the case, it would not be unreasonable, of course, to release the glider.

Glider Out of Position

Glider pilots practice 'boxing the slipstream' as a training exercise. It should not be carried out near the ground. Although uncomfortable for the tug pilot, it shouldn't normally cause control difficulties. If the glider gets way out of position near the ground and this leads to the potential for handling difficulties, consider releasing the glider and saving yourself and the tug! As a general rule, if you need full deflection of the elevator, it is time to let the glider push the tow-rope from his/her end – in other words, release the rope from the tug immediately!

Request for Towing Aircraft to Turn (USA)

In the USA, the glider pilot can fly out to one side thereby pulling the towing aircraft's tail out, requesting a turn in the direction the tug is subsequently pointing.

Too Fast or Too Slow Signals (USA)

It's interesting to note that in the USA they have the glider roll its wings to indicate a request for more speed and yaw to request less speed.

The 'old' winch launch signal in the UK for too slow used to be rolling of the glider's wings – but pilots were spinning in from low speed, so it was changed to the safer signal of lowering the glider's nose. This is obviously inappropriate for aerotowing.

Towing Aircraft Unable to Release (USA)

In the USA, the tow aircraft yawing (as opposed to waggling of rudder, which indicates a problem at the glider end, like airbrakes open) means towing aircraft cannot release. If the glider is not able to release *and* the tug is not able to release it's really not your day!

Tug Upsets

Tug upsets occur when the glider pilot gets too high and lifts the tug's tail uncontrollably. This tends *not* to happen from a pilot flying consistently high on tow, but rather from a pilot in difficulties a little low, perhaps in the slipstream, who suddenly 'winches' up behind the tug. This creates so much lift, and therefore drag on the glider, that the tug is not only tipped but also loses its forward momentum.

From time to time over the years, tug upsets have occurred at low level from which the tug has been unable to recover, usually with fatal results. A glider pilots' aerotow training emphasizes that correct position behind the tug is essential and that they must release if they lose control, get too high, or lose sight of the tug (See slipstream illustration). However, tug pilots must be vigilant during the early stages of the launch for any tendency of the tug to be pitched nosed down. At all times, monitor the tug's attitude, and if a significant backpressure is required to prevent any nose down pitch, release immediately. Be aware that tug upsets can happen rapidly and with little warning, and the release loads can be higher than normal.

There are a number of factors that increase the possibility of a tug upset:

- A glider that is to be towed from a belly hook
- Gliders with high-set wings relative to the towing hook
- Gliders with a low wing loading, usually older or vintage types
- The presence of turbulent conditions, especially if associated with a strong wind gradient
- Glider pilots with low hours and/or aero-tow experience
- Light-weight pilots
- The use of short towropes will exacerbate the problem
- *The list is not exhaustive, and we are always learning*

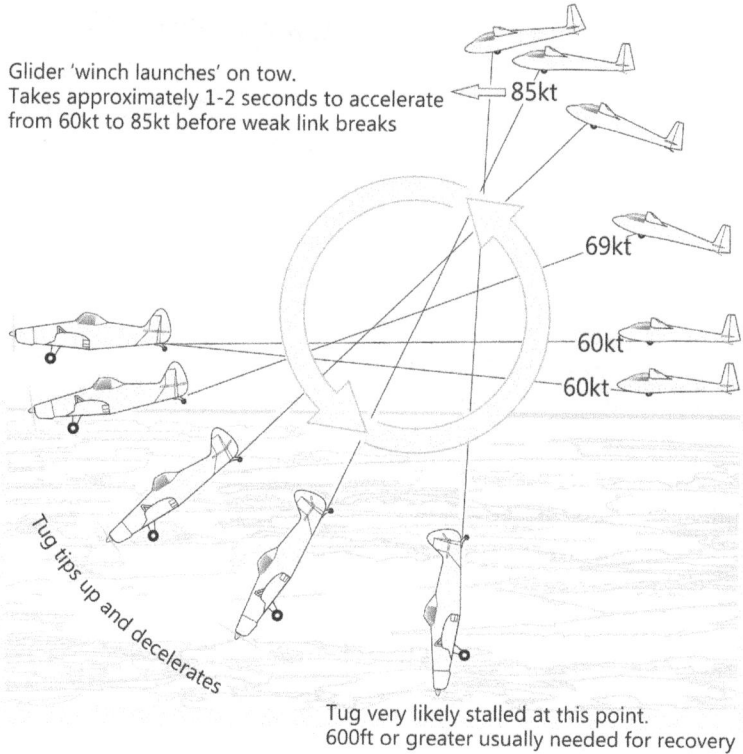

Glider 'winch launches' on tow.
Takes approximately 1-2 seconds to accelerate from 60kt to 85kt before weak link breaks

85kt

69kt

60kt

60kt

Tug tips up and decelerates

Tug very likely stalled at this point.
600ft or greater usually needed for recovery

A typical sequence is shown in the illustration. In reality, the situation is worse than shown because, as the glider zoom climbs behind the tug, its total energy increases (simultaneous increase in height and speed). This energy can only come from the momentum of the tug and therefore its speed will rapidly decay. This means that just when a high down-load is required to be generated by the tailplane/elevator to retain control and break the weak link on the rope, its capability to do so is vastly reduced by the decay in airspeed. This may result in the tailplane, and possibly the wing, stalling.

Typically, 600 ft (200 m) or more may be required to recover from an upset.

Also, it is important to avoid a hasty transition from level acceleration to climb, as this will result in the glider becoming low relative to the tug. This can tempt the glider pilot to attempt a rapid recovery, with obvious potential for over-correction.

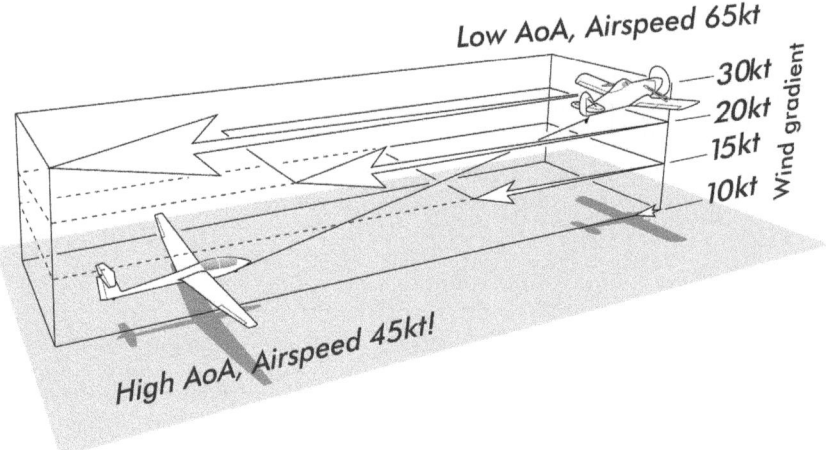

Tug pilots should avoid doing anything just after take-off, except fly the aircraft. Retracting flaps if fitted should be left until a safe height, say at least 300 ft (100 m).

Another cause of tug upsets occurs when a glider pilot performs a climbing turn on release before confirming that the rope has been released. Arguably, this is not as dangerous because it is normally performed much higher above the ground, but it could still give the tug pilot quite a fright. There are other destabilizing influences for both tug and glider pilot, such as re-trimming, flap and undercarriage retraction, instrument scan, or a canopy coming open. In a tug upset condition, bear in mind that the rope-release pressure can increase significantly with some release-hook designs.

Sometimes, the upset occurs so rapidly that the tug pilot has no chance to react and release the glider. If any glider pilots gives cause for concern, do not hesitate to release the glider before they can jeopardize the tug and pilot. Be sure to advise the duty instructor or person in charge afterwards so that further training can be arranged. To promote an effective safety culture it's important that this re-training should not be considered a punishment.

At the aero tow release height the student inadvertently pulled the trimmer fully aft instead of pulling the release causing the glider to zoom up into a tug upset situation. Fortunately the instructor pulled the release in good time and that combined with plentiful height saved the day. A thorough flight examination using various scenarios was carried out by the club concerned.

- *The instructor was* 'relaxed' *but hand was placed* loosely *behind and close to the control stick. The resultant pitch up was 30 degrees plus – uncontrollable, tug upset*
- *The instructor was* 'relaxed' *but hand was placed* firmly *behind and close to the control stick. The resultant pitch up was 20 degrees plus – probable tug upset*
- *The instructor was* 'standing by', *expecting the control stick to come backwards. This stopped the pitch up, the glider was re-trimmed – no tug upset*

The conclusion of this club's well prepared and executed assessment was that to prevent catastrophic tug upsets at low level it is recommended that the instructor flies with hand, not on the control stick but forming a definite 'block' close behind it, on standby to reverse any unexpected pitch up.

Lateral Tug Upsets

Another, dare we say, lesser danger to the tug is the situation leading to a lateral upset. This is a result of the glider going out to one side and progressively diverging until the tug reaches its control limits. If the tug pilot continues to apply full rudder it is possible to stall the tugs fin as the angle of attack increases. The sudden loss of directional control at this point can be spectacular and very close to a flick manoeuvre. The violent yaw is caused by the rudder no longer opposing the rope tension, therefore allowing the glider to pull the tail round. As a result of the rate of yaw the secondary effect in roll is also very significant and can go beyond vertical.

The need to release immediately is obvious because if the glider remains attached, the vertical upset scenario could develop.

The lateral upset can be avoided by using caution when applying large rudder deflections. If more than, say, half rudder is insufficient to prevent further yaw, be very careful and allow the tug to yaw slightly. If there is a significant increase in rudder load or the glider continues to diverge, release. If the rope is released or the weak link breaks while full rudder is applied, the sudden yaw can also be alarming but not as violent as a fin stall. The tug will yaw towards the glider, presenting a collision risk. The highest risk of a lateral upset is during a demonstration of the 'glider cannot release signal' used in many countries. As this involves a heavy, two-seat glider going a long way out of position; it should only be demonstrated with sufficient height.

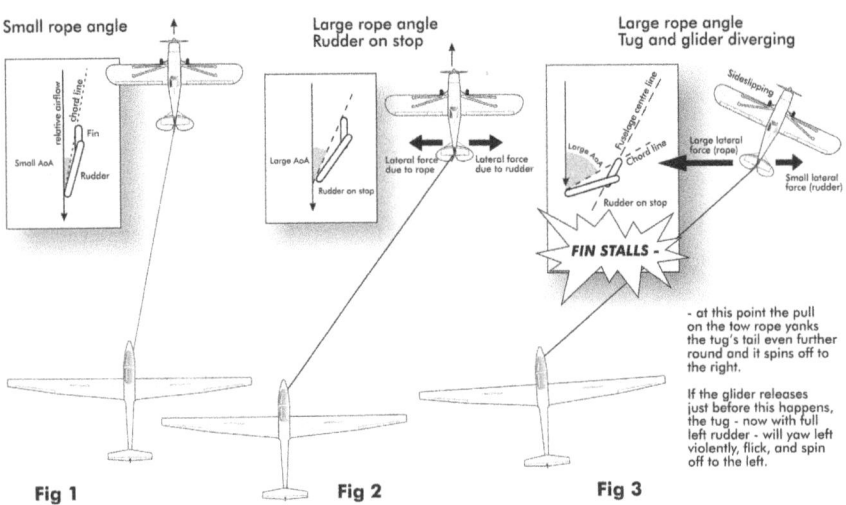

Glider Unable to Release

If in a two-seater glider, the pilot should try the other release first. Please note that there have been instances of the rope getting tangled around the glider's main wheel, nose wheel, or skid. In this situation the glider pilot will probably not be able to release. If you are in radio contact, talk to the tug to confirm that you can't release. In many countries, the glider pilot unable to communicate by radio would fly well out to the left of the tug (to a position where the tug pilot possibly sitting in the left seat, can see you) and rock their wings positively from side to side. Take care not to confuse this with a wobbly student practising out of position exercises. The glider pilot should first rock left and furthest out, or you'll end up swinging back towards the middle. While out to the left, the glider pilot may need a small amount of airbrake to keep the rope tight.

No immediate action is required. Normal practice, in the UK anyway, would be to tow the glider back towards the airfield and a suitable position. If necessary, manoeuvre to put the glider slightly higher than the tug aircraft and slacken the rope, so that when the tug pilot releases their end of the rope there's less chance of it smashing the glider's canopy or tangling around the glider. If the rope is taut, reduce power a little and check in the mirror that the glider is slightly high before releasing.

In some countries, the glider pilot will follow the towing aircraft through to a landing on tow. A French military gliding club insisted we do this once when we were visiting glider pilots. It's quite a good handling exercise, but you would be unlucky if both the glider and tug couldn't release. If the tug aircraft released the rope at his end, it would normally trail the glider at an angle of about 45 degrees below the glider. The glider pilot would normally plan a deep landing with the rope trailing. There could, of course, be a problem if the glider belly or winch hook was used for aerotowing. The tug released rope could come off the glider back-release mechanism and possibly fall on someone or something, so in this case think about where to release. Once again, radio communications will help enormously.

Tug Emergencies

If you have problems in the tug, consider waving-off the glider. This should be a positive rolling manoeuvre using about 30 degrees of bank so that the glider pilot cannot mistake it for a general wobble or turbulence. If your problem is an engine failure, simultaneously lower the nose to maintain a safe flying speed – don't be so preoccupied with signalling the wave-off that you subsequently spin in!

In the event of a major problem, do not hesitate to release the glider. The time taken to give the wave-off may compromise the tug's safety.

Consider all options. For example, if you believe that the emergency would benefit the utilization of an alternate airfield for landing, use the options open to you.

Tug Aircraft Engine Overheating

If the Cylinder Head Temperature (CHT) is approaching the red line, consider accelerating slightly, as long as it is of no consequence to the glider. Also think about reducing power and accepting a lesser rate of climb. Do you have cowl flaps you can open? If all this fails, wave the glider off in a safe position and land.

Total Power Loss

Release the glider immediately, simultaneously lowering the nose and trim for the aircrafts best glide speed. Carry out the standard forced-landing procedure. Statistically, most engine failures don't occur with a bang; they tend to be more gradual, so watch out for the signs.

Some tug pilots advocate turning through 90 degrees over the upwind boundary as a standard departure. This of course is if not geographically or operationally restricted, and at a safe height. The advantage to this is that if you are unlucky enough to have an engine failure or other extreme emergency after you have made this initial turn, you and the glider pilot might be better placed for a landing back on your airfield:

As the club's flight-safety officer, I received the following email from a club member:

I can add a further account, which reinforces a recent Bicester airfield (UK) experience, where a tug had a partial engine failure shortly after getting airborne. A friend was towing in a Gipsy-engined Chipmunk at Dishforth, Yorkshire, a number of years ago. He didn't experience a complete engine failure, but due to a sudden fault, it ran down to about 1,100 RPM as he went across the hedge with a Swallow glider on the back. Fortunately, he had carried out the 'ninety degree turn across the boundary' manoeuvre, and so it was a relatively simple matter for him to dump the glider and rope (without a signal, of course), and then make a landing on a cross-wind runway. Naturally, the Swallow was better placed energy wise and was able to do a similar manoeuvre onto the grass in the middle of the airfield. The wind was comparatively light and Dishforth is a large airfield, but nevertheless the 'ninety' made the whole thing comparatively easy and avoided a much more interesting manoeuvre into a field.

I have long been a strong advocate of the 'ninety degree turn across the boundary' both when towing and also in motorgliders. It's not my idea; I was taught it by Derek Piggott a number of years ago at Lasham (I am reluctant to say how many) when he cleared me to instruct in the original Motorfalke that had the most appalling rate of climb. It's also not always suitable because of local geography; however, most of the time it can successfully be used to increase your options.

I suppose the principal message is assess the conditions on the day (wind) then keep looking at fields on the way out and stay flexible generating as many options as you can.

Martin Durham

Serious Engine Vibration

A rough-running engine is sometimes a symptom of carburettor icing. Failing this, check the mixture and magnetos, and try changing tanks if you have the option.

Shed propeller tips can cause serious vibration. To prevent the engine shaking from its mountings, shut the engine down, slow as to stop the propeller, and carry out a forced landing. A wind-milling propeller creates more drag than a stopped one. If in a critical phase of flight, consider keeping the engine running at minimum power to save the situation and ensure a safe landing.

ASI Failure

This could be caused by pitot blockage. Use pitot covers to protect your pitot statics but don't forget to remove them before flight. Normally, check that the ASI is increasing during the ground roll and consider aborting the tow before getting airborne if the ASI is not increasing, if practicable and if safe to do so. If you do get airborne, fly by attitude and consider climbing to the top of the tow as long as you and the glider pilot are happy. Radio communications here could help with fine selection of speed. If you are unhappy to continue the tow, wave the glider off near the airfield and conduct a normal-powered approach using your familiar attitudes and power settings.

Brake Failure

If there is a risk of collision during taxi due to brake failure, turn the engine off using the magnetos. This will stop the engine immediately so that if you hit something you hit it without the propeller turning. If you are taxiing on a hard surface, consider turning onto grass to aid stopping. If you have a brake failure after landing on a limiting strip, consider going-around and subsequently diverting to a longer and/or more into wind runway.

Weather Difficulties

Poor Visibility or Low Cloud

If you are ever unhappy with the weather conditions, do not let a glider pilot, duty instructor, or gliding club official pressure you into giving a launch. The average private pilot with no real instrument flying experience will probably lose control in a matter of seconds once visual reference has been lost – even with a full instrument flying panel. Anyway, glider tugs are often not equipped for instrument flight. It is possible for a tug pilot with the necessary instrumentation to cloud fly and the glider pilot to remain 'in formation'. However, I think it's generally accepted it's best to remain clear of cloud, even if a lowering cloud base on a cross-country trip means a field landing or diversion is required. At high latitudes in the winter, the sun barely gets above the horizon. I've seen pilots get airborne in visibility that initially looks okay, to then turn into sun to then see absolutely nothing.

If caught by a local deterioration, for example a large shower, consider holding off until it has cleared. Alternatively, consider diverting to a neighbouring airfield and having a cup of coffee until it has cleared through. If landing in heavy rain, the windscreen could become obscured and visual landing clues become distorted. Look out sideways to judge your height and land well into the airfield to avoid obstacles but avoiding an overrun.

Strong winds

Strong winds will inevitably be present for glider pilots wishing to ridge or wave fly. The good news is that a strong headwind will reduce your take-off and landing run.

What is considered a strong wind?

As general guidance the UK CAA have recommended the following in one of their General Aviation Safety Information Leaflets (GASILs, which are available online):

"We continue to recommend that if the wind strength is greater than 2/3 of your aircraft's stalling speed, or ½ the stalling speed of a tailwheel aircraft, the aircraft should remain tied down or in the hangar".

Now for the bad news:
- Beware of turbulence and curl over from nearby trees, structures, or hills or from mountain wave-induced rotor (foehn effect)
- Exercise caution in the initial climb, as a strong wind gradient increases the tug-upset risk (as described in tug upsets)
- Check the position of the glider after take-off to ensure you don't leave it behind in the ground effect – accept an increase in speed if necessary

It's very easy, especially in high-powered tugs, to end up climbing through 100 ft (30 m) with say 75 knots (139 kph) indicated while the glider is at 50 ft (15 m) and at about 55 knots (102 kph) having insufficient energy to catch up (see illustration). It is vital to hold the attitude and not chase the Air Speed Indicator (ASI), accepting a higher than normal airspeed through the gradient until the glider is stable behind the tug. Being towed slightly too fast is not as bad as falling into the slipstream, nearly stalled at 75 ft (20 m) with the tug climbing away from you.

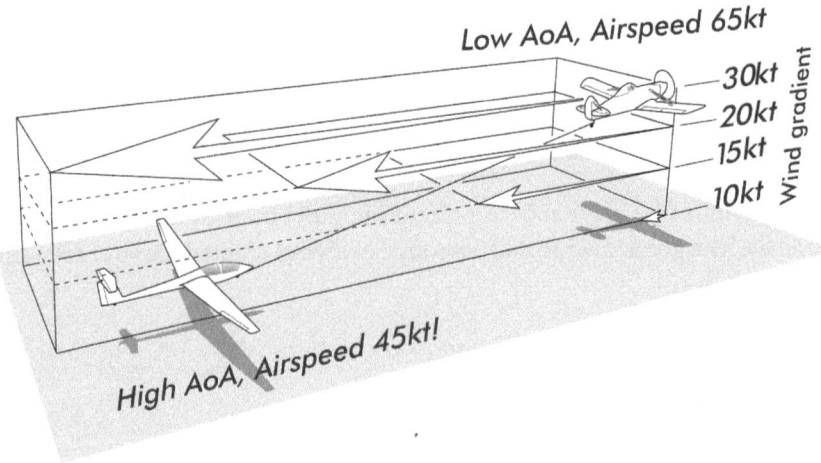

Low AoA, Airspeed 65kt

30kt
20kt
15kt
10kt

Wind gradient

High AoA, Airspeed 45kt!

Cross-winds and Tailwinds

The way to tackle cross-wind take-offs and landings can vary depending on type: high-wing, low-wing, flaps, no flaps, nose-wheel, tail-wheel, etc. Then there are differing techniques: side-slipping approach, crabbing approach, etc. Some techniques are recommended in the pilot's notes and some are just personal choice.

In my experiences with cross-winds, it's often easier to operate off a grass runway as opposed to a hard runway – given the choice, a flat grass strip is more forgiving.

The glider will often get airborne before the tug. The glider pilot may attempt to maintain position by using rudder, which will effectively result in a sideslip and slightly decreases take-off performance. In strong cross-winds, the glider may drift down-wind. This will exert a sideways force on the tug that will in turn try to weathercock the tug into wind. This may be beyond the ability of the tug pilot to counteract, and therefore the tug/glider combination may swing into wind before the tug leaves the ground.

How much cross-wind component? Use the clock-code, i.e. apply the cross-wind as minutes on a clock, so 15 degrees = 15 minutes:

- Runway heading is 270 degrees and wind is 300 degrees at 20 knots

Difference is 30 degrees applied to a clock face is half way round, therefore half the wind-speed, 10 knots is cross-wind component

- Runway heading is 270 degrees and wind is 315 degrees at 20 knots

Difference is 45 degrees applied to a clock face is three quarters of the way round, therefore three quarters of the wind-speed, 15 knots is cross-wind component

- Runway heading 270 degrees and wind is 330 degrees at 20 knots

Difference is 60 degrees applied to a clock face is all the way round, therefore consider it all 20 knots of cross-wind component

Easy isn't it?

If operating a flap-equipped aircraft in a cross-wind, would you use lots of flap, less flap, or no flap? I was involved in an interesting instructor debate about this once. There are various advantages and disadvantages for using flaps. Much of it depends upon type. I've said this before, but it's important, so I will say it again – seek the correct advice from your specific pilot's notes or a pilot with loads of experience on type.

Very generally for cross-wind take-offs and landings:

- Be prepared for turbulence and wind gradient
- Coarse use of rudder to keep straight, especially at low speed
- Consider differential braking if you have it, to help keep straight – be careful as it could decrease your take-off performance
- Apply lots of aileron into wind to prevent the into-wind wing from lifting
- Operating motorgliders with outriggers on grass – I've found that applying aileron to keep the down-wind wing low and therefore outrigger trailing on the grass, will assist in keeping straight. This is easier after landing but for take-off the upwind wing can lift at rotation, requiring quick opposite aileron application to prevent the upwind wing from lifting – tricky!

During a flying-instructor renewal test once, I was asked if I was landing on a runway with a 90 degree cross-wind, would I choose to land with the cross-wind from the left-hand side or the right-hand side?

And the answer is … I'll take the 90 degree cross-wind from the left-hand side, please.

But why, I hear you ask? Well, if the cross-wind is from the left, any gusts will tend to cause the wind to veer to a more into-wind component (in the northern hemisphere anyway). This is due to diurnal variation.

An additional consideration when landing with a cross-wind is if there is also a tailwind component present. If you allow a swing to develop into the cross-wind a tailwind will tend to exacerbate the swing.

A further problem associated with these wind conditions is that the inertial forces are significantly greater as they increase in proportion with the square of the groundspeed. However, the fin/rudder effectiveness, being a function of airspeed, doesn't increase. These characteristics are particularly important on tailwheel types.

If landing with a tailwind, the rudder may need to be reversed at low rollout speeds. Consider landing with a direct 10-knot (20kph) tailwind. As you slow through a groundspeed of 5 knots (10kph), the actual airflow over the flying controls becomes a 5-knot (10kph) tailwind. This can cause the rudder to reverse in effect. This condition is also particularly important on tailwheel types.

Take-off and land into wind as much as possible and be prepared for any wind gradient. Do not hesitate to go around if badly rolled by gusts, speed excursion, or any other instability. Once you've landed, avoid taxiing or attempting to turn down-wind, particularly in light-weight tail-dragger types. If necessary, shut down or get someone to hold the tail and wings as you taxi. When taxiing down-wind, consider doing so on a hard surface rather than grass, as less power will be required to keep moving, but avoid heavy braking. On two seaters, consider strapping someone in the rear seat to move the centre of gravity rearwards.

Here is a report from the UK's Confidential Reporting Programme 'CHIRP' (Autumn 2010). I thank the pilot who has submitted a very honest and open report and have personally learnt from it:

WING DROP vs PROPELLER TORQUE

Report Text: *The task was glider towing and the constraints on this particular day were that the tow rope had to be dropped prior to landing and the only available landing runway had a cross-wind from the left of 15-20Kts. Also, strong thermal activity was causing surface wind variations with turbulence from adjacent trees affecting the final approach. Conditions were challenging but within normal operating limits. Of the 19 sorties flown by me that day only this one landing gave rise to serious difficulties.*

Just as the aircraft was about to touch down on three points, a gust lifted the port wing against full aileron deflection. I applied full power and the aircraft suddenly and immediately levelled. I was able to climb away with no further roll control problem.

From my youth I was aware of accidents to RAF Balliol aircraft when a student bounced and applied full power causing the aircraft to roll as a reaction to propeller torque. The 235 hp Pawnee has a relatively high power weight ratio and while I have never previously experienced any effects of propeller reaction, in normal operation I apply or reduce the power slowly.

My conclusion is that the sudden and welcome correcting roll was a reaction to propeller torque, a second and unpleasant thought is that if the cross wind had been from the other side, the aircraft might have rolled inverted.

Lessons Learned: *Pilots of powerful light aircraft should be aware of the potential effects of propeller torque. The aircraft can roll against the direction of rotation of the propeller with a rapid application of power.*

CHIRP Comment: *The torque effects from a rapid application of power on large engines are well known to pilots with experience of single engine 'warbirds' and similar larger types. If you should experience a rapid wing drop, applying rudder in the same direction as the aileron input should assist in counteracting the uncommanded roll.*

In a situation such as that described, it is important that the appropriate corrective action has been thought out in advance so that it becomes an instinctive reaction.

When taxiing (especially in a tail-dragger types):

- Cross-wind from ahead and to one side – normally hold ailerons into cross-winds and elevator up
- Cross-wind from behind and to one side – normally hold the ailerons out of wind and elevator down or neutral

Or, for simplicity:

- Climb into wind
- Dive away from wind

Section 3 – Advanced Procedures

Glider Retrieves from Airfields, Strips, or Fields

Before embarking on a retrieve it is important that the following points are considered:

- The opportunity for accidents or incidents are greatly increased
- Radio communications between towing aircraft and glider are a distinct advantage
- Aircraft specific insurance might be required
- Permission must be obtained from the airfield operator or landowner
- Tug aircraft should normally be refuelled before departure – performance permitting
- The tug pilot should book out and let others know the plan
- A spare rope could be carried (consider a shorter rope to assist getting out of a field)
- A suitable map and/or GPS must be carried
- Consider which aircraft to use. A Pawnee or similar type is probably most suitable because of its good take off and climb performance
- Don't normally land away from your airfield with the tow rope still attached – it could snag on something/someone and it increases landing distance required
- If radio procedures are required, and many airfields do, the pilot must also hold a radio licence
- Be aware of the logging and pricing procedures
- Allow plenty of time – something will inevitably delay you, it might get dark, or you might run up to local closing time or curfews
- Take a mobile phone (cell phone)
- Plan thoroughly

Many clubs will restrict off-home-airfield retrieves to more experienced pilots as it certainly requires skill, currency, and the ability to think outside the box.

For a field retrieve, on arrival overhead or in the circuit take some time to assess the landing area for approaches, size, and surface. Gather as much information from the landed glider pilot as possible. If you are not sure of a safe retrieve, return to your club.

If at an active airfield, ask the locals what run will be most suitable for departure, considering surface, length, wind, slope, animal stock, climb-out, and other traffic requirements. Many airfields will not appreciate a glider blocking their main runway for long. It is unlikely that you will get the standard aerotow signals from ground helpers. Consider hooking on the glider yourself before starting engines, and then the glider pilot should close the airbrakes when fully ready to launch or make some other similar arrangement.

Before take-off ensure that you do the following:

- Work out your contingency plan to cover actions in the event of an aborted take-off, or rope break, or release soon after take-off. Brief the glider pilot and the wing runner accordingly
- Decide on a tow speed to suit glider and pilot
- Decide how the signalling is to be arranged and any other relevant details
- Be sure you have checked performance to get airborne and clear obstacles (See information in performance section about performance factoring)

As previously mentioned, it is beneficial to have radio communication between tug and glider pilots whenever possible.

After take-off, fly an obstacle clearance climb out to a safe height before accelerating to your agreed towing speed. Once a safe height has been reached or at a chosen altitude, reduce power to achieve a slight climb or level flight. If possible, it is preferable to reach the top of the climb just as the glider comes within gliding range of home as a cruise climb makes it easier for the glider pilot to keep the rope tight but mind you don't climb into controlled airspace.

Try to avoid the need to descend on tow, as it can be quite destabilizing. If a descent is needed because of cloud or airspace, reduce power slightly to achieve a smooth, shallow rate of descent. Anything more will require the glider pilot to use airbrakes. Glider airbrakes can be stiff to operate at high aerotow speeds, particularly if worn. High, left-hand loads will often translate into inadvertent right-hand movement in pitch on the control column.

Surging is when the towrope slackens and tightens continually. It can be caused by turbulence, lack of concentration on the part of the glider pilot, or by descending too quickly or inadvertently. It appears to be worse on a long tow-rope. Surging can occur during level flight, but is most likely during descending flight. The best solution could be to apply power and accelerate or climb slightly until the surging ceases and then slowly return to the desired stable state. The glider pilot would have the option of increasing drag by opening the airbrakes, lowering flaps and/or gear, but operating levers can induce a destabilizing pitch with the other hand, so maybe best avoided.

Remember that, until the glider releases, you are in charge of both aircraft and responsible for navigation, collision avoidance, avoiding entry into cloud, and normally any radio calls.

I landed at a big but disused airfield trailing a rope to tow out the second of two gliders that had unexpectedly landed there. I approached over one solitary post that later appeared to be the last remaining post of an ex-fence. I couldn't see this post from the approach. You know what I'm about to say – the rope snagged the post, and the weak link broke. I had considered flying over the field and dropping the rope but it appeared to be such a big, clear area that it didn't seem necessary.

Dual Tows

Dual tows are not recommended for normal club-towing operations as there is no clear advantage in doing them; in fact, there are additional hazards. Having said that, they can be useful for glider retrieves from big field or airfields. They are quite good fun and can be a useful training exercise.

Dual-Tow Performance

Consider the performance implications of towing more than one glider. Is your tug capable, authorised, and insured to do this? (See section on performance factoring).

Wind and Turbulence

Avoid cross-winds, strong winds, or turbulence. In this case it would be best to fly another day.

Dual-Tow Ropes

A special long and short dual towrope is required, with weak links at each end to protect both the gliders and the tug. Naturally, the weak link at the tug end needs to be stronger than those at the glider ends.

Positioning on the Ground

My advice is to lay out the ropes on the ground at an angle of about 30 degrees either side of the tug. Once the gliders are hooked up, check that the long rope is clear of the forward glider. Position the gliders to minimize any slack in the ropes.

If towing dissimilar types, it could be easier to put the glider with the lowest wing-loading on the short rope and on the down-wind side, as it will get airborne first.

Pilots

The pilot flying on the long rope is responsible for ensuring that his rope stays clear of the other glider, but it's normally a bit trickier to maintain station for the pilot flying on a short rope.

Radio

On a dual tow, if all three machines have radio, use a common frequency, note call signs, and complete a radio check before take-off. Any problems can then be called on the radio but bear in mind that use of radio for an inexperienced glider pilot under a high workload may be a complicating factor. Also consider using formation call sign when talking to Air Traffic Control (ATC) agencies.

On a cross-country flight, it's normal for the tug pilot to control frequency changes, but it is useful for the gliders to acknowledge each change and then check in on the new frequency. It's a good idea to brief and note the en-route frequencies before departure and designate a fall-back frequency if a pilot miss-selects and gets frequency lost or needs to chat. It is very easy for any one of the combination to call 'Dual tow – go company' or something similar in order to sort out any problems. The international allocated frequency for exchanging operational information is 123.45. The alternative is for the gliders to stay on a familiar gliding frequency and the tug pilot can report any ATC clearances that affect the progress of the flight.

Dual-Tow Signalling

Conventionally, you need three ground signallers: one for each glider (who also may hold the wings level) and a forward signaller who is the 'master signaller' and can be clearly seen by the tug pilot. The forward signaller can be dispensed with when signals are passed by radio, which is probably better. The short-rope wingtip holder and signaller (if different) have to be on the outside wingtip for safety. As the tug moves slowly forward, the glider signallers signal normally and give 'all out' once their rope is tight, but the forward signaller gives the tug pilot 'all out' only when he sees *both* glider signallers giving 'all out'. In a large field, if you can't find a third signaller, the more experienced glider pilot on the long rope can transmit 'all out' to the tug by radio. This may be better in any case in an off gliding field situation if the wing holders and signallers are not gliding people.

Consider this – if all the slack has been removed by previous positioning of the gliders, obviously there will be no slack so the tug pilot can go straight to all-out. Maybe this would be simpler and safer?

Positioning on Tow

Firstly, notice the terms I have used here. The glider position just above the slipstream is known as 'normal tow' as opposed to 'high tow', which is an undesirable position because of it possibly contributing to a tug upset.

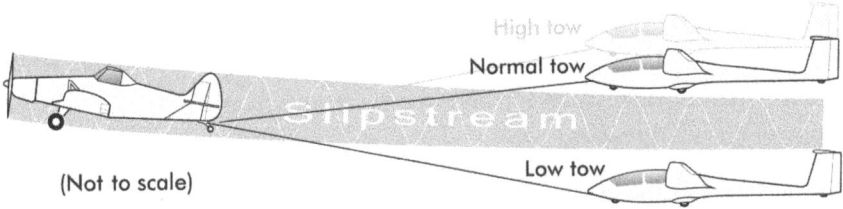

(Not to scale)

Some have found it successful to have the gliders go into normal and low tow behind the tug, short-rope high and long-rope low. The advantage is that, for the short-rope glider, the tow is virtually normal and requires no new flying techniques such as flying out to the right for long periods. Also, the low tow is a comfortable position for long cross-country tows. Regardless of the method chosen, it is vital that the pilots involved are fully briefed on the procedure that is going to be used.

Some have found that a bit of lateral and vertical displacement was most comfortable. That is gliders displaced in height with normal and low-tow positions and laterally by about a wingspan.

In stable air, the separation can be reduced to maximize efficiency. If it is turbulent or whilst manoeuvring, increased separation reduces the chance of the long rope touching the forward glider.

It has been noticed that if the low-tow glider gets too low, it can slow down the whole combination quite markedly – even with a relatively high-powered tug.

After Take-off

After take-off, the gliders initially retain their small lateral offsets from the tug. The tug should continue straight ahead without turning. At an agreed safety height, usually about 300 ft (100 m) above ground level (agl), the forward glider gradually goes across into the normal tow position behind the tug. Using this as a cue, the long-rope glider then goes into low tow in line behind the tug. Only when the gliders are in line behind the tug should it turn. The tug pilot should roll into all turns as gradually as possible, and use less bank than usual, to make it easier for those behind the tug. Even if they are experienced, pilots may be tired at the end of the day after a long soaring flight, and that is when dual-retrieve tows can be quite common.

Release

At the release point, the forward glider releases first and does a positive climbing turn so that the tug pilot can see in his mirror that he has gone. If the rearmost glider releases first there is a danger of the long rope snagging around the foremost glider. The lower glider then comes into normal tow (as opposed to the low-tow position) so that the tug pilot can see him, and releases with another positive climbing turn, normally turning the opposite way from the first glider, unless the pilot can see that he is well clear.

After Release

The slower the tug descends after release, the less chance there is of the ropes tangling. Remember that if the rope combination is long, you might need a larger height margin than normal to clear obstacles on the approach.

General

Dual tows usually go well while the tug is climbing, but most glider pilots have little experience of long, level, cross-country tows, especially in the low-tow position. More dual instruction on the use of airbrakes on tow and descending on tow will be an advantage. It may well be worth setting up a training flight involving circling the airfield level and slightly descending at a safe height before letting some pilots go on a dual tow away from the airfield.

Whilst in level flight, it is worth considering towing with the wheels extended, as the extra drag could help stabilize the tow. Similarly, it is worth briefing the possible need for airbrakes at anytime. The rear glider has the advantage as they can see the whole combination and can react quickly. It is less destabilizing to select a small amount of airbrake early, rather than a large amount at the last minute.

Tug performance must also be carefully considered. Expect a very long ground roll and climb rates of potentially a lot less than the normal performance with a single glider on tow.

Let me bore you with my war story!

Back in the early 1970s I undertook a dual tow. I was on the long rope in an ASW15, the glider on the short rope was a K6CR glider and we took off from runway 24 at Bicester. As we crossed the airfield perimeter, the rope became detached from the tug.

I had just moved into line with the tug and below its slipstream. For a few seconds I was distracted by the rope and the three foot wooden spreader bar coming towards me. (Fortunately they did not wrap round the glider.) I lost sight of and forgot about the other glider, which turned through 180 degrees and landed down-wind back on the airfield.

So there I was, not very high, and the fields ahead did not look landable. I could not turn right because of the RAF domestic site, so I turned left, made a snap field selection and landed in the field next to the Buckingham Road. Fortunately, I stopped just before hitting the far hedge, but then had to fend off a herd of cows.

What worries me in retrospect was that I had completely forgotten about the other glider. It would have been interesting to say the least had we both chosen to land in the same field, not to mention the collision risk.

The dual tow rope was, I believe, never recovered. It fell into the small sewage works which used to be located roughly where the Bicester ring road now meets the roundabout. The field I landed in was long ago built over.

Tim Harrington

Glider Aerobatic Towing

This is to describe the requirements of the aerobatic glider pilot and to see how the tug pilot can help maximize the practice time in the air, from a competition perspective.

The following types of flight might be applicable:

- Students being taught dual aerobatics
- Solo/dual practice for a full competition sequence
- Practicing specific aerobatic figures
- Competitions

We can also add club spin training, but that's not specifically a concern here, though some aspects will apply. Also, we are not thinking about going inverted on tow either!

A good start would be a thorough pre-flight briefing so that everyone knows what they are supposed to be doing. Radio contact with all concerned could also yield significant flight-safety benefits.

High-performance aerobatic gliders normally require towing at a higher than normal speed – check with the aerobatic pilot.

A competition aerobatic flight uses some line-ground feature (e.g. runway, railway line, etc.) as the into wind axis (even if the wind is up to 45 degrees off this). On the ground is a 1 km by 1 km area with the top of the 'box' at 4,000 feet above ground level. The tug runs the glider in at typically 4,000 feet above ground level, into wind in the middle of the box. Usually just before the centre of the box, the glider releases and starts the sequence.

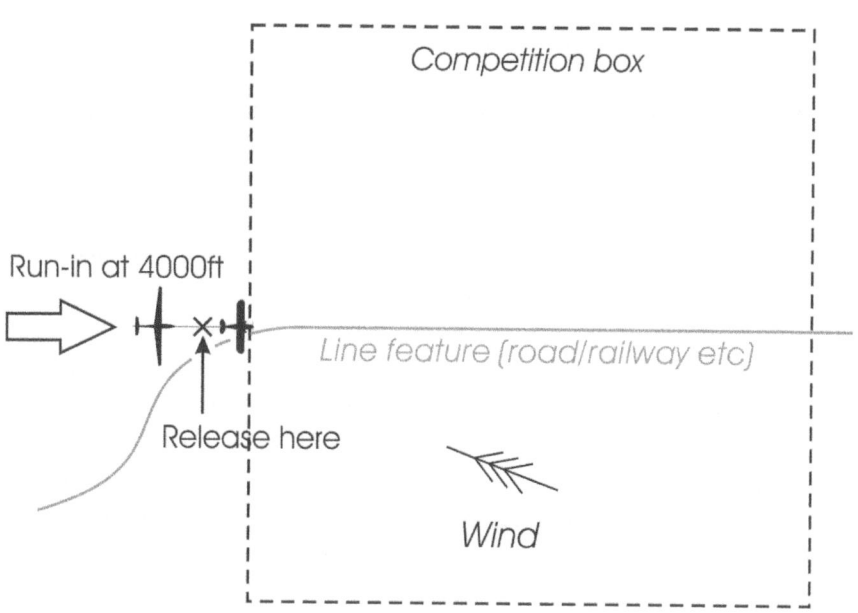

Competition box

Run-in at 4000ft

Line feature (road/railway etc)

Release here

Wind

What would be very useful to the glider pilot is that:

- The tug becomes stabilized on the run in for a few seconds with wings level at the agreed release height and on the agreed heading (into wind) so the glider pilot can start thinking about his sequence and where to release
- If the tug pilot accelerates on the run in, the glider can gain a couple of hundred feet in the pull-up
- After release, the tug then needs to disappear from the aerobatic box really quickly – it must not stay in the box any longer than necessary and obviously must not fly beneath the glider

For different wind directions, it would be good if both the glider pilot and the tug pilot agreed some standard drop-off points. It's important that the glider aerobatic pilot can:

- Do the full sequence without overflying the circuit or flying anywhere there could be traffic conflicts
- Get back to the airfield easily
- Because it is quite common for the glider to beat the tug down, watch out for a glider in the circuit.

Glider Cross-Country Competition Towing

As with many things, this probably starts with a good pre-flight briefing by the person in charge of the competition towing:

- Are you and your tug prepared?
- Know what you as a tug pilot are supposed to be doing
- Know what the glider pilots are supposed to be doing
- Know the relevant radio frequencies and complete a radio check
- Know the glider drop-off points
- Know the take-off and landing areas
- Think 'what-if' scenarios
- Maintain a really good lookout
- Consider manoeuvring lots in the descent to make yourself more conspicuous in crowded airspace

Things that could be different to a normal club towing situation:

- Times of intense activity
- Gliders carrying water ballast that might require a few extra knots airspeed and additional take-off performance considerations
- Take-off and landing areas might differ
- Different radio frequencies might be in operation
- Competition gliders will normally be dropped in one specified location
- The normal signal for 'release immediately', positively rocking of the tug's wings, is sometimes used for "I've towed you to the competition start height in the drop-off area; it's time to release"

To summarize, keep a really good lookout, maintain situational awareness, and condition yourself to expect the unexpected.

Section 4 - Engine Management

Aerotowing gliders will put extra strain on a tug aircraft's engine so it was considered that an expanded section on engine handling could be beneficial to the aerotow pilot.

The following advice is very general and is based on the average club tug, (if there is such a thing). However, it is vitally important to operate your particular aircraft in accordance with the manufacturer's or pilot's notes for that type, including any towing supplemental data sheets.

I thought the best way to approach the somewhat complicated issue of engine management was to list things in an engine start order and to 'go round the cockpit', discussing the various engine controls.

Let's start with the **Electrical Master Switch**, which isn't strictly part of the engine at all. It does, however, normally need to be on to allow starter engagement, which is why it's mentioned here.

The **Beacon or Anti-Collision light (ACL)** does pull some battery power,. The advantage is that, when selected 'on' before engine start, it conventionally warns 'outsiders' of impending engine start and the dangers associated with that. If your aircraft has high-intensity strobes, these are best left off until taxiing onto the runway as they can be quite distracting to others (especially at night but then again we don't normally tug at night). I recommended that strobes are also left off in the vicinity of refuelling areas because they produce high-energy electricity that could ignite fuel vapour.

The **Fuel Selector** handle is often quite removed from the fuel cock itself, so it might be unwise to totally trust the handle or pointer. Try to 'feel for a detent' when you move the handle. Try selecting the fuel off (on the ground of course) and time how long it takes for the engine to stop. You might be surprised to see how long it takes. Turning the fuel off occasionally to stop the engine verifies its operation. It's often not part of any maintenance schedule to check this, and I've had an engine just keep running despite supposedly turning the fuel cock off .The other advantage to occasionally turning the fuel cock off is to exercise it, but do remember to turn it back on. (See advice earlier, in reference to the Pawnee's fuel cock systems).

Clubs have different philosophies regards turning the fuel off overnight in the hangar or leaving it on. I've mentioned this before, but as it's important,

I will reiterate. My view is that it best left on. If it's regularly turned off, sooner or later someone, especially in a tugging environment, will try to take off with it turned off. If no engine run-up checks have been carried out, often the case when carrying out aerotowing – the engine will likely stop at a very inopportune moment!

A **Primer** is not always fitted, especially on injected engines. Most are just a syringe and squirt fuel directly into the intake manifold. They are normally separate from the aircraft fuel system apart from the fuel being drawn through the fuel selector. When using the primer, ensure the fuel is first turned on, pull it out, *and wait for it to fill with fuel* before pushing it back in. Make sure it is locked after use as it can, and has, caused an engine to stop when you least want it to. How many times do you use it? Seldom or not at all on a warm engine; more on a cold one but be careful not to over-prime. When an engine produces a few short bursts of black smoke on start and then dies, the engine is probably flooded. To cure this problem, have a coffee and then see if fuel stops dripping from the induction system and try again. Alternatively, set the mixture control to cut-off, open the throttle fully and attempt a start. When engine fires immediately, retard the throttle and advance the mixture, but to do this properly you really need three hands! This system should work for both carburetted and injected engines alike. Obviously, attempting a start with full throttle selected can be potentially dangerous. Make sure the aircraft brakes are set, wheels chocked, and a competent person is at the controls.

My Fournier RF4 has no starter motor, therefore if I suspect flooding, I ensure the magneto is off, open the throttle and turn the propeller backwards a few turns to 'blow out' the excess fuel – remembering to reset the throttle before attempting another start.

The **Fuel Boost Pump** is often fitted to low fuel tank aircraft. It is normally put on before engine start; this enables you to check it works by seeing a rise on fuel pressure gauge and often a clicking noise is heard. It also primes the fuel line and helps suppress vapour formation that can cause vapour lock, which in turn can cause an engine to stop. The fuel-boost pump is often turned off after engine start and left off for taxi-out. This tests the correct operation of the other, mechanical fuel pump. The fuel boost pump should be selected back on before take-off. Vapour formation is often associated with hot conditions and with the use of **mogas**, which

114

is simply motor car petrol that is occasionally authorized for aircraft use. There are, as always, advantages and problems associated with using mogas. For example, at the time of publication, the UK Light Aircraft Association (LAA) recommends a maximum fuel temperature of 20 degrees centigrade, a maximum operating altitude of 6,000 feet and that the aircraft is not left for extended periods of time in direct sunlight. The advantage is that it is cheaper than avgas. Check before using it.

The **Starter** control comes in various guises. In fact, on some Robin types it is cleverly hidden under the fuel-off cock, preventing you from starting the engine with the fuel turned off. Common variations include the rotating key that goes off/right/left/both/start. The start position often causes some confusion, as some require a push *and* turn to engage the starter.

Another way of doing it is to have a key (or switches) to select the **Magnetos** and a separate button to engage the starter. If your machine has the ability to select the magnetos separately, it's normal practice to select the impulse magneto, often the left one, for start (See section on General Engine Operations). The impulse magneto retards the timing to assist starting. The non-impulse magneto is not retarded, so it may cause the engine to kick back and has potential to damage the engine. Once the engine is running, select both to prevent the spark plugs that aren't firing from fouling-up. It's probably worth reinforcing the fact that a magneto is a self-exciting generator and is independent of the aircraft's electrical system. It defaults to live – that means that when turned off it is earthed by a lead normally known as the P lead. This lead is prone to breaking, which would make the magneto *live* – which in turn means the engine could fire if turned! There are three ways of checking if you have a live magneto:

- The first method is a 'drop not stop' check carried out after start or before putting to bed for the night and carried out at low RPM. As it says on the tin, you are looking for a slight drop in RPM without a stop of RPM. Select each magneto in turn. A stop would indicate an unserviceable magneto, and no drop would indicate a live magneto
- The second method is during the magneto checks carried out on the engine run-up. If you have a live magneto there will be no RPM drop when turning that magneto off. When selecting each magneto in turn, be careful not to turn both magnetos off at fairly high-power settings then immediately back on again as this can damage the engine. This is also a 'drop not stop' check but carried out at run-up power. The problem with completing this check is that in an aerotowing environment it is often only carried out once a day on the first flight
- The third method is the 'dead-cut' check which could be conducted when shutting down the engine to ensure it is safe to be put in the hangar. You will now be checking to ensure that neither of the magnetos is live when the ignition switch is selected off. To conduct this check at idle RPM both magneto switches should be moved momentarily to off and then back to both. When 'off' is selected the engine should falter as if it is about to stop, proving neither magneto is still live. Some people reckon this check can damage the engine and prefer the 'slight drop not stop' check at idle power.

I will leave it up to you to decide which method you think is best.

Starter motors can burn out. This can be expensive, so follow the published guidance for your aircraft to prevent it. As a very general rule, don't engage for more than 30 seconds and then include a 3-minute cooling period. As I'm sure you know, before starting the engine have a good lookout all around the aircraft – look for people in the vicinity of the propeller, animals anywhere that could be startled and run into the propeller, and people or equipment behind. When it is all clear call 'clear prop' *and then wait a moment.* This is often missed. Pilots make the call then immediately attempt a start not giving anyone you have not seen the chance to move clear of the propeller.

If you have a stack of avionics, they can be all turned on and off using the **Avionics Master Switch** rather than selecting each one in turn. It is normally off for start as the engine start can cause a potential current surge that can damage avionics.

The **Mixture Control** is often continuously variable between fully rich and cut-off, which sometimes is known as Idle Cut Off (ICO). This is the normal way to shut down most engines. Removing any fuel present in the engine has the advantage of reducing the chances of the engine firing (which it can do even with the magnetos off) and injuring someone with the propeller. Select ICO, wait for the engine to stop, and then immediately turn the magnetos off and remove any key.

I never touch a propeller unless I have to, and in many cases you simply don't have to. There could be times to turn the engine off with the magnetos. Brake failure while taxiing would be one such occasion. If you are likely to hit anything or anyone the quickest way of stopping a rotating propeller is by selecting the magnetos off. Bumping into something or someone with a stationary propeller is one thing, but with the propeller still turning is highly likely to be much worse.

The time from touchdown to shutdown can be critical for some engines. After taxi-in a brief cool-down period – at idle power and parked into wind – is often a good idea. This evens the Cylinder Head Temperatures (CHTs) and helps prevent the cylinders cracking. Select the boost pump and avionics off. After the engine has stopped the very next thing to do is to make it safe by turning the magnetos off, removing the key, and placing it on the instrument cowling or somewhere where it can be easily seen. I have seen a key-type magneto selector worn enough to allow the key to be removed with one of the magnetos still live. Needless to say, this is extremely dangerous, so simply seeing the keys removed might not necessarily mean the magnetos are safe.

There appears to be much conflicting advice on the use of the mixture control in tugging operations. In most cases, the general advice for glider towing is to leave it rich, possibly except for a very high-altitude tow or cross-country tow. When left fully rich, the excess fuel helps cool a hard-working engine towing one or more gliders. Adjusting it incorrectly can cause all sorts of damage to the engine. A slightly rich engine has no real problems but a slightly lean one does. The only exception I can justify is if the engine is so rich it begins to run rough, it is then permissible to lean it normally, even in the climb. As always, confirm these practices with the manufacturer's notes.

If you are operating from a hot and high airfield, you will probably need to use the mixture control in order to achieve maximum possible take-off power. This is sometimes done by selecting full power before take-off, holding on the brakes of course, and then leaning the mixture for peak RPM. Then, enrich to a point about half way between no leaning and best RPM. If you are fortunate enough to have an EGT (Exhaust Gas Temperature) gauge, do the same but lean to peak EGT, and then enrich about 100 degrees Fahrenheit. If you are lucky enough to have a multi-channel EGT gauge, do the same but reference to the leanest cylinder. Note, the leanest cylinder is not always the hottest but is the one that reaches peak EGT first as the mixture is leaned. I think most pilots would agree that having a very slightly rich mixture results in little extra fuel used, little loss of power, but greater cooling. If you have a constant speed propeller, normally do as above but lean to maximum manifold pressure. The process of leaning takes a few seconds, so take your time.

The **Throttle** controls the amount of air going into the engine, not directly the fuel. The carburettor or fuel injector responds by metering fuel by a comparative amount, so you are only indirectly controlling fuel flow with the throttle. It's the mixture control that has direct control over fuel flow. However, there is often a 'fuel squirter' that supplies a short burst of fuel into the carburettor when the throttle is rapidly opened. This 'accelerator pump' is also sometimes used to prime prior to engine start by repeatedly pumping the throttle.

Carburettor Heat is used to prevent carburettor icing. This can occur for three reasons:

- Firstly because of Boyles Law, which says that pressure and temperature will drop in a venturi, and a carburettor is basically a venturi
- Because of fuel evaporation
- Finally, impact ice that builds up on air intakes, filters, and alternate air valves can lead to restricted airflow.

Carburettor icing can happen almost anytime – it doesn't necessarily have to be cold because it's mainly to do with relative humidity.

BUILD-UP OF ICING IN INDUCTION SYSTEM

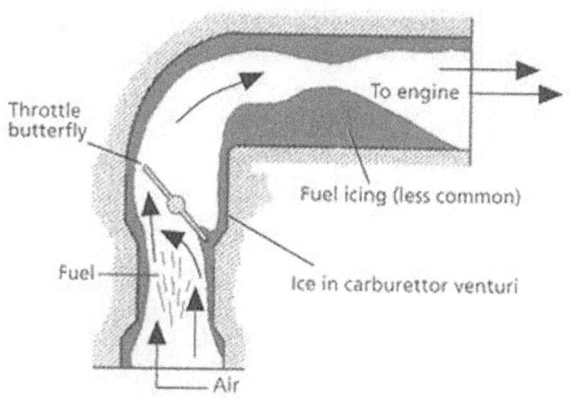

CAA Carto DO C(G)6 Drg No 8805b 23-11-84 10-5-90

Carburettor icing is more likely at low-power settings and less likely at high-power settings. It's not normal to select carburettor hot air in the climb as it can cause detonation, which is the cylinder firing before it is supposed to. During the cruise, you should use carburettor heat regularly – selecting it fully on (hot) for long enough to melt any carburettor icing – then select it off (cold). Prior to selecting the carburettor heat on, note the engine RPM. After selecting it off, check that RPM is the same as previously noted. A higher RPM indicates carburettor icing may have been present. We don't usually use part hot air unless the aircraft is fitted with a carburettor icing gauge, so normally it should be just on or off. With low power settings in the descent it is normally on (hot) all the time. Briefly revving the engine in the descent is also very desirable as warm air for the carburettor is drawn from around the exhaust exchanger.

This engine warming should be carried out regularly – say, at least every 30 seconds and not be based upon descent altitudes of 500 ft for example, as a machine with a good glide performance may take quite a long time to descend those 500 ft. Advancing the throttle in the descent has the added advantage of checking that the engine actually does respond. If it doesn't, at least you get to discover this higher up! Before beginning a descent with low-power setting, select carburettor hot air on first – this gives the engine a shot of warm air – *then* select a lower power setting. Carburettor hot air is normally unfiltered.

You will see from the diagram that severe carburettor icing is likely in temperatures of -2 to +24 degrees Centigrade at all power settings and moderate to high humidity.

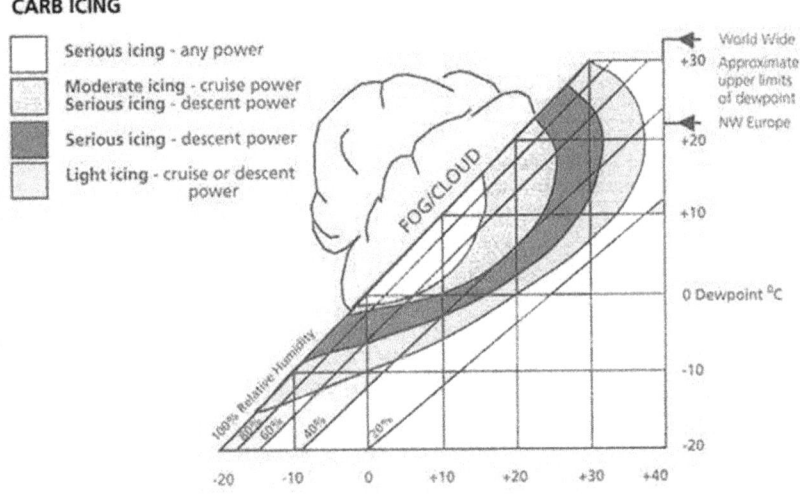

The **Propeller Pitch Control** is used to control propeller RPM; it's *not* an indication of power. That is normally displayed by a manifold-pressure gauge. On a hydraulic type system, it's normal procedure to cycle the propeller pitch a few times before take-off to replace cold oil in the prop dome with warm oil from the engine. It also confirms correct operation. Electrically operated systems are becoming more common, especially on motorgliders.

An **Alternate Air Control** is often fitted to an injected engine. Impact ice can form on the air scoop or filter. Alternate air feeds unfiltered air into the system, normally from inside the engine compartment.

Cowl Flaps are sometimes fitted. These are fairly common on some motorgliders, which are also sometimes used for glider towing. The cowl flap controls air out of, and therefore into the engine area and helps control Cylinder Head Temperature. As always, operate in accordance with the published information, but generally for towing operations, the flaps would be open to aid engine cooling in the climb. They might be closed in the descent.

The **Tachometer** is designed to display engine RPM. For a fixed-pitch propeller it is normally the primary source of indicated engine power. The **Hourmeter** normally converts revs to engine hours and is not purely engine hours running. It will tick over slowly at low RPM and faster at high RPM, so if you are just flying locally and don't need airspeed or power, it might be worth selecting a lower RPM in order to extend your servicing interval and to save fuel. The hourmeter reading is often used for engine servicing intervals.

Another way of measuring hours is the **Hobbs Meter**. This can be operated by oil pressure for engine running time or by air speed or a weight-on-wheels switch to indicate flight time. Sometimes the meter is directly wired to the master switch.

Manifold Pressure often accompanies a constant-speed propeller and is normally the primary source of indicated engine power. It indicates pressure in the intake manifold. The instrument is basically an aneroid barometer and is normally calibrated in inches of mercury. If today's pressure is 29.92 inches of mercury the manifold pressure should also read 29.92 – until engine start.

The **Oil Pressure Gauge** is often mechanical, involving no electrics, but not always. Pressure should come up to a specified value within a certain time after engine start. The oil pressure increases slowly on the first start on a cold day. Very generally, if there is no pressure after 30 seconds, shut down.

Oil Temperature may take some time to rise off the bottom stop because it often starts reading at relatively high temperature.

I've only one thing to say about **Fuel Gauges** – don't trust them! Always try alternative ways of verifying fuel on-board. Looking in the tank is often an easy way of doing this but not possible on some types. Some aircraft have an independent low-fuel warning light – don't totally rely on this either.

Exhaust Gas Temperature (EGT) is used for fuel leaning and general troubleshooting. Once again, as many are manufactured in the USA they are often calibrated in degrees Fahrenheit. They can be of the single or multi-probe types.

Cylinder Head Temperature is often calibrated in degrees Fahrenheit. It is sometimes only connected to one probe in a single cylinder – normally the hottest.

General Engine Operations

The impulse coupling is a mechanical gizmo contained in some magnetos. Its job is to improve starting by doing two things, retarding and increasing the intensity of the spark. Many aircraft only have one impulse magneto, normally the left, but some aircraft have two. You can often hear the impulse magneto making a snapping sound as you pull over a propeller. Safety - remember to treat propellers with great respect and always treat them as live.

Impulse couplings are great if they are looked after, but can cause engine failure if they go wrong. Ensure yours are regularly and properly maintained.

Some aircraft have a separate **Alternator Switch** – this is normally left off until after engine start as it pulls some valuable battery power for energizing alternator field windings and it also reduces the chances of electrical spiking during start.

Propeller swinging is best avoided. The very important safety considerations appear earlier in the book.

Flooded Starts were discussed with the mixture control.

Starting Problems – In cold conditions, consider pre-heating the engine and the battery. The battery could be removed and placed in a warm place.

It might be worthwhile turning the propeller over a few times (safety!). Some manufacturers recommend leaving the primer out, and pressing it in while starting the engine. If you are having trouble starting a warm engine, some manufacturers recommend having the carburettor heat on as it aids fuel vaporization and reduces the density of the incoming air – this artificially enriches the mixture, but remember that this air is normally unfiltered.

Misfiring is often the precursor to an engine fire. Carburettor fires can happen without the pilot's knowledge and can cause considerable damage. The standard advice for a carburettor fire is to keep turning the engine over in the hope of sucking the fire into the engine. I think it's reasonable to say that if you see flames, shut down, turn the main fuel cock off and get out quick! If you are parked on a flat surface where there is no chance of the aircraft moving away, consider releasing the parking brake. This will enable the aircraft to be pushed away from any burning fuel that may have spilt under the engine. Take the aircraft fire extinguisher if you can.

Engine Run-up Checks are conducted to confirm the engine is fit to fly. First, check that the brakes are set – lookout (including behind) – then, set the power recommended by the manufacturer. Don't stare at the power instruments, but initially look outside to ensure the aircraft is not creeping forward. The carburettor hot-air check is normally carried out first to ensure an ice-free carburettor for the other checks. Leave the carburettor hot air on long enough to melt any ice present. The carburettor heat cold stop should be reached before the lever is fully home to ensure the valve is fully closed. Hot air leaking into the carburettor could significantly reduce take-off power. Indications of carburettor ice could include rough running at high power setting as the ice melts, and then the RPM sitting at a higher setting after selecting carburettor heat back to off, indicating the ice has melted during the application of the hot air.

The **magneto check** – Different aircraft manufacturers recommend different RPM drops – even for the same type of magneto. It's not really important, but what is important is to check that each magneto can power the engine at reasonable power settings and that there is a slight drop in RPM or power from switching each magneto off in turn. If there isn't you might have a live magneto as previously explained. Don't leave only one magneto selected for too long or the other plugs might foul-up. Go to both magnetos between each individual magneto check and be careful not to inadvertently turn both magnetos off then on again as this could damage the engine.

Listen to the engine throughout – it will tell you lots about its health. If the RPM drop on a single magneto is accompanied by rough running, a plug could be fouled. A general fix could be to try the check again at higher RPM. If that doesn't work, try it again with the mixture about one third lean (for about ten seconds).

Some tug pilots like to **sideslip** in the descent. The advantages of this could be increased rate of descent with less rapid cooling of the engine, and increased conspicuity as the aircraft descends sideways. The disadvantage could be high bending loads on the propeller and crankshaft. As always, it could be type specific.

During the **climb out**, it might be necessary to reduce RPM or power to stay within maximum continuous (max con) limits. It's not normally important to rush the power reduction; in fact, some statistics suggest it is the power reduction that can cause engine problems to arise. I suggest you make any adjustments at a height that gives you some options if the engine does misbehave.

Engine Emergencies

Firstly, about one in seven light-aircraft accidents are caused by engine malfunctions not associated with fuel starvation.

If you have an engine malfunction:

- First – Fly the aircraft
- Second – Fly the aircraft
- Third – Fly the aircraft

You get the idea? After that consider:

- Applying carburettor hot or alternate air if fitted
- Checking both magnetos selected
- Change fuel tanks if possible
- Checking the engine instruments
- Reduce power if possible

- Checking the mixture
- Checking the primer is locked
- Land at nearest suitable place
- **But not necessarily in that order**

Early application of carburettor hot air might melt any ice present and enable a restart.

Engines often show signs preceding a failure – look out for them.

As always, your pilot's notes should give you more specific information.

Engine Fire – Select main fuel cock off and cabin heaters off. Consider letting the engine run until the line fuel has been used-up. Also consider side-slipping to help smoke and flames clear the aircraft and to aid lookout. Close the cowl flap if fitted to reduce air available to feed the fire and land as soon as possible putting out a 'Mayday' call. After landing, evacuate taking the fire extinguisher if possible.

Low Oil Pressure is normally accompanied by high oil temperature (if not, it's possible a gauge problem). Reducing power or increasing speed will often help. If not, land as soon as possible.

High Oil Temperature is normally accompanied by low oil pressure as described above. Consider increasing speed and/or reducing power.

High EGT – Reduce power, enrich mixture and increase speed if possible. If you have a cowl flap consider opening it.

Summary

As you might have surmised, much research and consultation has gone into the production of this book. I sincerely hope you have found it useful. If local laws and restrictions are appended here at the end, the book could constitute a concise guide to your operation. Please use the following space to make your own notes and let me know of any feedback at johnPmarriott@gmail.com

With your feedback I hope to produce a revised version, but that might have to wait until retirement!

About the Author

Having spent the major part of my gliding life at glider winch-launch sites. Compared with some I'm not actually that experienced a tug pilot. I have a few hundred hours mainly on Pawnees, Supermunks (Lycoming-engined Chipmunks), Robins, and a little in Super Cubs. However, I do have over twenty-five years of experience as a full-category gliding instructor, motor-glider instructor, light-aircraft flying instructor. I also have loads of free time (not!) and because of this the British Gliding Association in its wisdom asked me to be its senior glider-tug pilot.

I have acquired some 17,000 hours in powered aircraft, including over 3,000 in light aircraft and motorgliders, and over 1,000 in gliders. I pay the mortgage by being a Boeing 777 captain for British Airways. I also have a safety job for the British Airline Pilots Association (BALPA) and am an accredited accident investigator for the International Federation of Airline Pilots Association (IFALPA).

A few years ago, while I was a relatively bored first officer flying Boeing 747s and at the ripe old age of forty-four years, I embarked on a distance-learning MSc course in Air Safety Management. Thanks to six-day trips sitting around some remote destinations with not much else to do, I managed to complete the MSc two years later.

I'm very lucky, as gliding clubs in the UK are generally well run and have fine tugmasters who execute a sensible, pragmatic, and safe approach to their operations.

Over the past few years of doing the job, I've wrung the minds of many experienced tug pilots and tried to bring together best practices and disseminate them amongst our community.

Reader's Own Notes, Including Local Restrictions and Laws:

Lightning Source UK Ltd.
Milton Keynes UK
UKHW011917260122
397754UK00001B/317